Douglas Carnegie

Law and Theory in Chemistry

A Companion Book for Students

Douglas Carnegie

Law and Theory in Chemistry
A Companion Book for Students

ISBN/EAN: 9783744666572

Printed in Europe, USA, Canada, Australia, Japan

Cover: Foto ©Paul-Georg Meister /pixelio.de

More available books at **www.hansebooks.com**

LAW AND THEORY IN CHEMISTRY

A COMPANION BOOK FOR STUDENTS

BY

DOUGLAS CARNEGIE, M.A.

SOMETIME SCHOLAR AND DEMONSTRATOR IN CHEMISTRY OF GONVILLE
AND CAIUS COLLEGE, CAMBRIDGE

LONDON

LONGMANS, GREEN, AND CO.

AND NEW YORK: 15 EAST 16th STREET

1894

PREFACE

THIS book contains the substance of a summer-school course of eight lectures delivered before an audience of school teachers of elementary chemistry at Colorado College, Colorado Springs, U.S.A.

It is not therefore adapted to beginners, but it is hoped that it may be found useful by such students as wish to recapitulate and co-ordinate the more important principles of chemistry before proceeding to more detailed and advanced works.

The qualification *companion book* occurring on the title-page is used advisedly, for text-book continuity and completeness cannot be claimed for the matter of the following pages. The seven chapters are in reality short and independent essays on the subjects of which they severally treat.

In the choice of subject-matter I was chiefly influenced by the two following considerations. Firstly, I desired to treat of those subjects which in my opinion are essential to a liberal understanding of the science, but which are inadequately treated in, or altogether crowded out of, the current text-books of elementary chemistry. In the second place, I wished to restate and emphasise such points as in my experience present especial difficulties to the student.

Throughout, the attempt has been made to keep the information up to date, and to indicate, with due appreciation of perspective, the trend of modern research in its relation to the science as a whole. In this connection it is perhaps scarcely necessary to state that the hope was to excite interest rather than to satisfy it.

I would here express my sincerest thanks to my friend, Mr. Pattison Muir, for the very valuable help he has given me.

DOUGLAS CARNEGIE.

BLACKHEATH.

CONTENTS

vii

LAW AND THEORY IN CHEMISTRY

CHAPTER I.

ALCHEMY AND THE BIRTH OF SCIENTIFIC CHEMISTRY.

In the ancient world the only scientific pursuits which were considered reputable, and which led to fame or distinction, were politics, philosophy, and mathematics.

Dyeing, tanning, bronze founding, glass making, soap making are all industries which boast a remote antiquity, but the early practice of these arts was confined to the slaves, and the ignominy of attempting any adaptation of theory to practice never for a moment suggested itself to the noble-born.

But while the actual practice of what we now understand by the term commercial chemistry was in the hands of the uneducated, and possessed no literature, the philosophers theorised on the composition and nature of matter, and to trace these theories and their consequences is, as we shall find, to trace the early history of our science.

The early Greek philosophers of the seventh to the

A

fifth centuries B.C. belonged to one or other of two
schools known as the Ionic and the Eleatic schools
respectively. The adherents of the Ionic school
believed that in all the apparently diverse things by
which they were surrounded there existed a common
principle or substratum (ἀρχη), and they sought to
evolve order and simplicity out of the apparent
diversity in terms of this common principle. According as the attempt was made to investigate and detect
this principle by the observation of natural phenomena,
or by the unaided exercise of the mind, we find Thales
asserting water to be the substratum of all things, or
Pythagoras selecting number as the one real thing in
nature. The attitude of the Pythagorean branch of
the Ionic school is well illustrated by the following
quotation from Plato's seventh book of the " Republic " :
" We shall pursue astronomy with the help of problems
just as we pursue geometry, but if we wish to become
thoroughly acquainted with astronomy we will let the
heavenly bodies alone."

The disciples of the Eleatic school founded by
Xenophanes did not attempt to explain the material
universe, but they espoused a philosophy of an essentially negative character. Assuming unity and im-
mutability in the universe, without attempting to
further specify or explain either the one or the other,
they maintain that change and the apparent diversity
of things are inexplicable delusions of the senses.
Reason was made the only criterion of truth, and as
reason apparently led to results at variance with the

information conveyed by the senses,[1] the latter were regarded as fallacious.

To summarise the differences between the two schools briefly, we may say that while the Ionic school sought to explain the many in the one, the Eleatic school denied the real existence of the many and assumed the one.

The views of Empedocles (440 B.C.) do not, however, strictly conform to either of the two schools just described. Unlike the Eleatics, he aimed at a rational explanation of the diversity and change in the universe, and unlike the Ionic school this explanation was given in terms, not of one, but of four common principles. viz., earth, fire, air, and water. These substances, to which Empedocles assigned mythological names, were regarded as the constituents or elements out of which everything else was built up. In Empedocles' opinion, their association in different proportions sufficiently accounted for all the diversities presented by the world to the senses. But to account for the continual changes that are ever taking place in these diversities, Empedocles had to assume the existence of two other rival elements, viz., love and strife,[2] the former causing the attraction and combination of unlike elements, the latter causing their disruption. If we substitute the seventy odd elements of to-day's chemistry for the four Empedoclean elements first mentioned, and then the

[1] Zeno's well-known paradox of Achilles and the tortoise may here be instanced.

[2] We should be inclined to regard the "love" and "strife" of Empedocles rather in the light of powers or influences than as elements. But Burnet (*Early Greek Philosophy*) states that Empedocles himself regarded them as corporeal elements co-equal with the other four.

term affinity for the remaining "love" and "strife," we cannot but be struck with the close analogy which the views of this early Greek philosopher bear to those held in the nineteenth century concerning the composition of, and the changes undergone by, the material universe.

The views of Empedocles were modified, and in their modified form popularised by Aristotle (385-322 B.C.). The great Stagyrite regarded all substances as portions of one and the same fundamental matter, modified by the greater or less quantities of the four "elementary principles"—earth, fire, air, and water—which it had impressed on it. This conception of elementary principles has nothing in common either with our present conceptions of the elements or with the original views of Empedocles. It is one of those abstract ideas originally borrowed in all probability from India, where Buddha taught that in addition to the four principles already mentioned, other two—ether and consciousness —existed, and it is very difficult for us to replace our clear-cut notions of definite concrete elements by these unreal products of the Oriental imagination.

It must not be supposed that Aristotle intended to convey the idea that water, air, fire, and earth, as we now understand these terms in a purely materialistic sense, constituted the universe by their manifold combinations. The elementary principle of earth implied to the Aristotelian school merely the possession by the bodies supposed to contain it, of the properties of dryness and coldness. Similarly the principle of water implied coldness and moistness, the principle of fire, heat and dryness, and the principle of steam or air, heat and moistness.

Accordingly it was held that one specific kind of matter could be totally changed into another by merely altering the proportion in which the principles had been impressed on the fundamental matter of the first substance—by altering the ratio of its heat, coldness, dryness, and moistness. Thus by imparting more heat to water (fundamental matter especially richly impressed with the principle of water) it becomes steam or air. By taking away its moisture and cooling it, it becomes, as was then taught, rock crystal or petrified water,—fundamental matter especially rich in the principle of earth.

The important point to note is that the Aristotelian doctrine favoured the idea of a possible transmutation of any one substance into any other.

Experiments having as their aim the change of one substance into another seem first to have been extensively practised in Egypt, which was once called *Khmi* (Greek, χημία) on account of the darkness of its soil; *Khmi* literally meaning black soil. And it came about that the art which had as its aim and object the transmutation of material things received the name of the country where it first flourished—the art was called χημία,[1] just as a certain form of wool-spinning is called worsted-making from the village of Worsted in Norfolkshire, where the industry had birth.

[1] The use of this term to designate the art first appears, so far as we know, in the fourth century.

In the seventh century the great wave of Islamism rolled westward from Arabia towards Europe, and the missionaries of Mahomet overran and conquered Egypt.

During their sojourn in Egypt, the Arabians found the dwellers in the land executing what were then regarded as actual transformations of matter.[1] For instance, when molten sulphur was poured into mercury, the metallic properties of the latter disappeared and the mercury and sulphur changed into a solid substance black as the raven's wing. When this black substance was heated gently for some time, it changed into the beautiful red substance now used as a paint under the name vermilion.

This does not strike us now as a very wonderful change, but owing to the way in which the early naturalists were wont to symbolise things, it was to them a change of the profoundest type. For black was with them emblematic of evil, while red denoted virtue; and the change from a black substance into a brilliant red one was in their eyes as the miracle of an evil tree bringing forth good fruit.

But the change which appealed most to the Arabian conquerors was the, to their minds, less miraculous change or transmutation of the base or common metals into the noble metals, gold and silver. Though the

---

[1] It is probable that prior to their invasion of Egypt, the Arabians had already learnt something of Egyptian arts through the Nestorians. In the fifth century this Christian sect was banished from its home in Constantinople to the desert of Thebais in Egypt, whence its followers gradually migrated eastwards.

modus operandi of this transmutation was kept very secret,[1] yet the possibility of the transmutation was regarded as a well-ascertained fact during the first centuries of the Christian era. The Arabians carried away with them in their western trend this idea of the ennobling of the base metals, and introduced it into Europe under the title *Alchema;* the prefix *Al* being the Arabic for the definite article, and *chema* being the Arabic rendering of χημία, but signifying dark in the sense of obscure or secret rather than black. In all probability the prefix *Al* was added for the purpose of conferring a dignity and distinction on the art as being the highest and most reputable of all arts, just as the Spanish discoverers of the New World called the American crocodile *el lagarto—the* lizard—because it was the largest and most formidable lizard-like animal they had ever seen. El lagarto and alchema persist in the language of to-day as the words alligator and alchemy. The word alchemy gave rise by apherisis to the word chemist, and this in turn to the word chemistry.[2] (Cf. poet and poetry.)

[1] The earliest chemistry of all is to be found in the ἄγια τέχνη or holy art, whose secrets were kept sacred under "pain of the peach tree" (*i.e.,* under the penalty of being poisoned by hydrocyanic acid), whose study was confined to the priesthood and the sons of kings, and whose laboratories were the temples.

[2] There are some who, rejecting the derivation of the word Alchemy given in the text, would derive χημία from χυμεία (a mingling, or infusion) from χυμός (juice), which in turn is related to χέειν (to pour). Hence the two current methods of spelling the word *chemistry = chymistry.*

Picton in his book *The Story of Chemistry* rejects both these deriva-

The Arabian alchemists, as we learn from the writings of Geber—works dating from the eighth century and constituting the oldest chemical literature extant[1]— had a theory of their own respecting the nature of the metals. Aristotle's theory was wide and applied to all kinds of matter, whereas the alchemistic theory was specialised and had reference more particularly to the metals. The latter theory, moreover, did not replace the former, but only supplemented it. Aristotle's elementary principles were still recognised, but regarded as the more remote factors in determining the properties of matter, while Geber's elements represented the more proximate constituents of matter in general and of the metals in particular.

Geber's idea, which met with a general adoption by the alchemists, was that the metals consist of sulphur and mercury in varying proportions. His sulphur and mercury did not, however, mean what these words now connote to us—definite chemical individuals with invariable properties. The names had for him an

tions, and asserts that Alexander of Aphrodisia (a great exponent of the Aristotelian doctrines in the second century A.D.) *invented* the word *chymike* for the operations of the laboratory ; but surely it is unusual to invent words out of one's inner consciousness without any attempt at appropriateness, historical or otherwise, in the resulting invention. Though Alexander may have thrown the word *chymike* current on the world, it does not follow that in the selection of the word he was uninfluenced by anything save inventive power.

[1] Some philologists would derive the word Gibberish [said to have been originally Geberish] from the alchemist Geber, notwithstanding the fact that Geber's writings, compared with the inflated and unintelligible jargon of other alchemical productions, are models of conciseness and clearness.

abstract meaning; his sulphur and mercury varied in kind and in properties. Thus gold and silver were very rich in "pure or perfect kinds" of mercury, but the one contained a red sulphur, and the other a white. In short, Geber's elements were rather of the nature of *elementary principles* than of the nature of *elements* as we now understand the latter term.

Since the change in purity as well as the change in relative proportion of these two metallic constituents was assumed to be under the control of the experimenter, it is obvious that the attempt to transmute the base metals into gold was quite legitimate from a theoretical standpoint.

Further, it was generally believed that the evolution of the base metals into the noble ones, gold and silver, was a change which was proceeding spontaneously and constantly in nature, and that therefore the alchemists' task virtually consisted in an artificial acceleration of a perfectly natural process.

Alchemy scarcely merits the dignity of being classed as a science. For some of its followers it was a transcendental philosophy which amounted almost to a religion; but for the majority it was nothing more nor less than an empiric trade. Its chief service to science lies in the fact that it stimulated experiment—blind and thoroughly haphazard experiment perhaps, but still experiment. Its most fatiguing and laborious study was prosecuted by the majority, not so much out of a desire for scientific triumph or for truth's

sake, as out of a thirst for that power which the so-
called philosopher's stone with its Midas touch was to
confer. The alchemists never dreamed of trying to
convert gold into lead.

The philosopher's stone was the name given to a
mythical theurgic powder [1] which was to have the
power of fermenting millions of times its own weight
of fused base metal into gold. The strangest and
most diverse ingredients were mixed together in the
attempts to prepare this much sought for powder;
we read of snails' slime, serpents' teeth, gall stones
taken from cats, blood, hair, white of egg, &c., in
addition to the never failing ingredient, mercury.
The philosopher's stone was generally identified with
what in more general terms was called the "One
Thing"—the perfect form of matter which was to
combine in itself all the properties of *all* other kinds
of matter in their highest perfection. Hence we can
in a measure account for the many and varied ingredients which were believed to be necessary to its
production.

Many of the receipts for making the philosopher's
stone that have come down to us are perfectly unintelligible, so allegorical is the language in which
they are couched. Sometimes the directions seem
fairly explicit, till one comes to the concluding item
—"add carefully a sufficiency of *you know what.*"

---

[1] It may here be stated that some philologists find in the generally
entertained belief that the "stone" would turn out to be a *black*
powder, a derivation of the word alchemy. See article "Chemie" in
Ladenburg's *Handwörterbuch der Chemie.*

Several reasons may be assigned for this secrecy and mystery, and sometimes one, sometimes another of these reasons may have been operative. In the first place, the alchemists were undoubtedly desirous of keeping their profession strictly limited as to numbers, a result which would be effected by the adoption of what practically amounted to a private code. Again, there were alchemists whose attitudes towards their profession are sufficiently expressed in the following quotation :—

"If thou shouldst reveal that in a few words which God hath been forming a long time, thou shouldest be condemned on the great day of judgment as a traytor to the majestie of God."

Finally, there may have been some alchemists who out of vanity or for purposes of self-aggrandisement asserted, with intent to deceive, their knowledge of the transmuting substance. Obviously it would be expedient for such men to express themselves so obscurely as to render it impossible for others to bring their statements to the test of experiment.

With the Benedictine monk, Basil Valentine, begins the period of iatro—or medical—chemistry, a period lasting from the fifteenth to the seventeenth century, during which transmutation was in abeyance. Heretofore the apothecaries had prescribed purely vegetable preparations only, but now we find mineral specifics contesting the field with them and partially replacing them. The true goal of alchemy had come to be looked on as the attainment of health, not

wealth—"a back tough as Hercules" rather than
the riches of Solomon—and the philosopher's stone
was endued with all the attributes of an elixir of
life in addition to its now secondary transmuting
powers.[1]   It was even customary at this period to
ascribe the great ages of the patriarchs to their
possession of the health-giving stone.[2]

It is probable that the idea of the prophylactic
and life-sustaining function of the stone originated
in the great propensity of the alchemists for symbolism and metaphor.   They even went the length
of constructing their apparatus symbolically—retorts,
for instance, being frequently fashioned so as to
resemble men, the bulb of the retort being the
stomach, the crown of the retort being the head,
and the beak being the nose.[3]

[1] Van Helmont (1577–1644), the originator of the generic term *gas*
and the discoverer of the existence of several distinct kinds of gases,
believed that the universal solvent or "alkahest," not the philosopher's
stone, would prove to be the true elixir vitæ.

[2] See Jonson's *Alchemist*, act ii. sc. i., where reference is also made
to the belief that such mythological stories as those of the Golden
Fleece, the Hesperian Gardens, the Boon of Midas, &c., were "all
abstract riddles of the stone."   It may be asked how it came about
that the patriarchs died, seeing that they were possessed of the elixir of
life.   The alchemist would have made answer in this wise.   Every
person has a predestined maximum lease of life—a limit which he
cannot possibly pass, but a limit to which by reason of disease he
seldom attains.   Now it was not claimed for the magic stone that it
was more mighty than destiny.   All it could do would be, by prevent-
ing sickness and consequent premature death at three-score years and
ten or thereabouts, to enable men to enjoy the maximum lease of life
stretching well up into the hundreds.

[3] See Bolton, *Trans. New York Acad. Sc.*, December 1882, March
1883.

In further conformity with this propensity, the alchemists also referred to the base metals as the diseased metals, the lepers, &c., and figuratively spoke of their transmutation into gold as a healing of their diseases. Thus in all probability arose the idea of the alchemist's elixir of life.

In the blind search for this elixir many valuable drugs were discovered and fairly fully investigated. Valentine busied himself chiefly with the therapeutics of antimony and its compounds, and he very thinly veiled his impatience with and contempt for "the deplorable, putrid, and stinking bag of worms," that failed to see the wonderful virtues of antimony as he himself saw them.

In addition to his book, *The Triumphal Chariot of Antimony*, Valentine left behind him a complication in the views which had been current up to his day respecting the nature of the metals. This complication, which consisted in the assertion of the presence in the metals of a third elementary principle, viz., salt,[1] did not inaugurate any material advance in views respecting the composition of matter. This new idea, however, was soon generally accepted, and in contemporary literature we find all kinds of fanciful analogies instituted between such diversities as the Trinity, body, soul, and spirit, and the trio salt $\ominus$, sulphur φ, and mercury φ.

---

[1] The principle of salt was represented in that portion of the metal which resisted the action of heat, and remained behind as a solid residuum.

Valentine had shown a laudable boldness and independence in ushering in, against strenuous opposition, the reign of iatro-chemistry, but his immediate successor, Philippus Aureolus Theophrastus Bombastus Paracelsus von Hohenheim—that strange character, half scientist, half charlatan—carried on the crusade with a vehement intrepidity amounting almost to truculence.   After forcibly expressing his contempt for the old school of "medicasters" by a public holocaust of the works of the celebrated physicians Galen (second century) and Avicenna (tenth century), he continued to work in the field first opened up by Valentine, and enriched medicine by the knowledge he contributed thereto of many valuable mineral specifics [1] (corrosive sublimate, tincture of perchloride of iron, &c.), which still figure in the pharmacopœias of to-day.   He also substituted the more or less pure active principles of plants for the crude electuaries of the apothecary, and it is interesting to note that it was probably through his introduction from the East, and liberal use, not of a mineral, but of a vegetable drug, viz., laudandum,[2] that Paracelsus effected those wonderful cures which gained for him his great and widespread reputation.[3]

[1] "The medicines are ranged in boxes according to their natures, whether chymical or galenical preparations."—Quoted in Johnson's Dictionary.

[2] Laudandum (of which the modern word for opium, viz., laudanum, is a contraction) is the Latin gerundive signifying "meet to be praised."

[3] In his poem "Paracelsus," Browning, while making use of incidents in the life of the chemist as a *mise en scene*, sketches the career of an ideal Paracelsus.

Early in his career, Paracelsus devoted himself to an experimental study of transmutation, but it was not long before he concluded that it was a chimera.   Some of the best minds among his contemporaries and im- mediate successors were doubtless influenced by the conclusions of so great a personality, and renouncing alchemy, devoted themselves to medical chemistry.   At any rate the decline of alchemy, properly so-called, dates from the time of Paracelsus.   Most of the so- called alchemists of the latter portion of the middle ages were in reality frauds of the lowest order, whose stock-in-trade consisted in a little *leger de main*, and an unlimited amount of the inflated and highly allegorical language in which the alchemists were wont to give their recipes for the elusive stone, and to impress the lay mind.

In the memoirs of the Academy of Sciences for 1772, Geoffrey makes an *exposé* of the ways and means of these pseudo-alchemists.   Hollow stirring rods con- taining gold, and temporarily stopped with wax, crucibles with false bottoms previously charged with the precious metals, and composite rods—one half gold soldered to the other half iron, and the whole coloured uniformly—were among the properties of the craft. Thus equipped, the charlatans practised on the cre- dulity of the public, easily leading it to believe them possessed of clairvoyance, and a supernatural power enabling them to control natural agencies.   Telling fortunes by the stars, dispensing "familiars to rifle with at horses and win cups," &c., they pursued a royal road to wealth. Instances of such frauds occur throughout mediæval literature and history. One of Ben Jonson's best constructed plays—*The Alchemist* —is based on the despicable chicanery of a "cunning man," well named Subtle, and his confederate, Face. In the *Canterbury Tales*, again, Chaucer, apparently moved by some sudden resentment, departs from the order of the poem set forth in the prologue to introduce "The Chanones Yemannes Tale." A yeoman or servant falls in with the body of pilgrims on their way to Canterbury, and, after detailing to them the seductive horrors of alchemy, relates the story of some trickery that his late master, the Chanon, had practised on a priest.

In actual history we find these pseudo-alchemists figuring in the retinues of bankrupt kings and spend-thrift potentates along with chaplains, jesters, and so forth. At the instigation of Edward II., we read that John Cremer, Abbot of Westminster, invited the famous Spanish *chrysopœus*,[1] Raymond Lully, to come over to England for the purpose of replenishing the royal exchequer; and Henry VI. for long enter-tained a firm and costly belief in alchemy, in spite of the fact that Henry IV. had passed a law—said to be the shortest in the statute-book—making the practice of alchemy a felony. Indeed, most of the European courts so fostered and encouraged alchemy during the

---

[1] Chrysopœia (literally, gold making) was a synonym of the term alchemy in use during a long period. See Ben Jonson's *Alchemist*, act ii. sc. i.

fourteenth and fifteenth centuries that the markets of the day were glutted with worthless and counterfeit coin.

Several coins and medals, which are claimed to be products of the alchemical art, are still preserved in the Royal Cabinet of Coins in Münich, and in the Imperial Cabinet in Vienna.[1] As many of these relics undoubtedly consist of pure gold and silver, one is forced to conclude either that the bullion from which they were made was produced fraudulently by such tricks as Geoffrey has explained (p. 15), or that the noble metals resulted from the application of cupellation processes to base metals which were believed to be pure, but which in reality carried noble metal.   In agreement with the first conclusion, we read that in 1709 a certain retainer alchemist, Domenico Manuel by name, being detected in his knavery, experienced the irony of execution on a gilded gallows at Küstrin.   In support of the alternative conclusion, it may be stated that there is in the British Museum a facsimile of a silver medal made by Becher (see *post*, p. 24), which was in all likelihood obtained by the cupellation of argentiferous lead supposed by Becher to be pure lead.[2]   The medal bears on its reverse the following inscription :—

"Anno 1675 mense Julio Ego J. J. Becher Docter hanc unciam argenti purissimi ex plumbo arte alchymica transmutavi."

[1] See H. C. Bolton's *Contributions of Alchemy to Numismatics;* also Reyher, *De numis quibusdem ex chymico metallo factis.*

[2] The original of this interesting medal is in the Vienna Cabinet of Coins.

On the obverse is the usual representation of Saturn
with his wooden leg, scythe, &c. For the alchemists
believed the metals to be under the influences of the
planets; and sluggish lead had assigned to it as its
patron star the planet Saturn—the slowest moving
planet with which the alchemists were acquainted.[1]

Becher, regarding the production of his medal as a
bonâ fide case of transmutation, seems to have valued
it simply as a chemical curiosity; for we are told that
he expressed himself more interested in the solution
of nature's riddles than in the heaping up of wealth.[2]

Even so late as the year 1843 we find one of the
pseudo-alchemistic cult, one François Cambriel, of 19
Judas Street, advertising in a leading paper his readiness to teach in a course of nineteen lessons all the
secrets of the Hermetic Art.[3]

But that the *bonâ fide* alchemists themselves thoroughly believed in the possibility of transmutation,
there can be no doubt.[4] Such facts as the following

[1] The personification of Saturn as a lame man with hour-glass and
scythe, &c., is due to the fanciful identification of the Roman Saturnus
with the Grecian Cronus during the Hellenising period.

[2] Becher's offer to the Government of Holland to provide it with
six millions of golden thalers *per annum* if it would provide a certain
amount of silver and unlimited sea-sand, does not seem to have been
taken up in spite of the fact that an experiment made in 1679 is said
to have turned out six times more productive than Becher had
anticipated!

[3] So called after the mythical Hermes Trismegistos, the reputed
founder of all arts and sciences, and the Grecian representative of the
old Egyptian godhead Thoth—the deified intellect.

[4] Even such reputed early-day chemists as Davy, Dumas, and
Bergman, could not bring themselves to utterly reject the idea of the
possibility of transmutation. Peter Woulfe, a fellow of the Royal

could not but foster the brightest hopes of men who, satisfied with surface views of things, never questioned what was under the veneer. Iron plunged into a green solution obtained by dissolving certain ores (containing copper) in nitric acid, was itself apparently changed into the more valuable metal copper ; and copper melted with tutty (an impure oxide of zinc) acquired the bright rich yellow colour of gold. And in the metallurgy of the alchemists much importance was ascribed to purely colour changes; much was supposed to have been achieved in making base metals take the colour of the noble ones. Indeed, many of the early-day alchemists seem to have believed that to give a base metal the colour and sheen of gold was in effect to change the base metal into gold. In accordance with these beliefs, the transmuting substance is often called a tincture (from *tingo*, I dye). Thus, on a medal of the year 1647 made by one Hofmann, are inscribed the letters T G V L, probably an abbreviation of the sentence *tincturæ guttæ V libram*, meaning that five drops of the tincture used had effected the transmutation of a pound of the base metal from which the medal had been prepared.

Again, in the process now known as cupellation, lead heated in a vessel made of bone ash, slowly disappeared and left behind a button of silver. Pyrites treated in a similar way often left a legacy of gold. Of course the noble metals were *ab initio* present in

Society of this century, whose name lives in connection with certain pieces of chemical apparatus, was a firm believer in transmutation.

the lead and the pyrites, but the experiments were for the alchemists, who had no notion of what we now understand by chemical homogeneity, cases of indubitable transmutation.

When one recalls the modern work on the marvellous effects of "traces of impurities" in altering the properties of large masses of mátter, one can scarcely be surprised at the attitude of the alchemists. Instance in this connection Carey Lea's so-called allotropic forms of silver. Carey Lea has recently prepared, among others, a blue pulverulent form of silver soluble in water, and an insoluble variety of the colour of burnished gold. These interesting bodies are not, it is true, pure silver; yet they contain some 98 per cent. of the precious metal, and would have been startling discoveries in the days when transmutation was regarded as a possibility. Reference should also be made here to the marvellous effects of great industrial significance produced in the properties of iron by the addition of mere traces of such substances as aluminium, tungsten, &c.

The transition from the vague and uncurbed fantasies of alchemy into the true science of chemistry, was marked by the appearance of a work written in the form of a discussion conducted by a symposium of scientific men, and entitled "The sceptical chymist, or chemico-physical doubts and paradoxes touching the experiments whereby vulgar spagyrists[1] are wont to

[1] Chemistry was often referred to as the spagyric art, from $\sigma\pi\acute{a}\omega =$ I separate, and $\acute{a}\gamma\acute{e}\iota\rho\omega = I$ unite.

endeavour to evince their salt sulphur and mercury to
be the true principles of things." In this work Robert
Boyle (1626–1691) strongly contests "the doctrines of
the four elements, and the three chymical principles
of mixed bodies." From the quotation just made, as
well as from the remarks of the debaters, it is clear
that a confusion of ideas had taken place with regard
to the Aristotelian philosophy, and that actual earth,
fire, air, and water, had come to be assumed by many
as being actually present in, and as forming real constituents of diverse forms of matter—there had been
in fact a resuscitation of the original Empedoclean
views.

Boyle, in language which knows nothing of the
mystic and rococo style of alchemical literature, up-
holds the modern view of elements, defining them as
certain primitive and simple bodies which, not being
made of other bodies or of one another, are the ingredients of which all those called perfectly mixed
bodies are immediately compounded, and into which
they are ultimately resolved.

It is clear that if gold and silver are elements in
this sense, i.e., simple undecomposable, indestructible
forms of matter created once for all in fixed and unalterable quantities by a special act of creation, then
all attempts at transmutation are necessarily in vain.
The steady growth, since Boyle's day, of the conviction
that gold and silver, and indeed all the metals, are
elements in this sense, has been contemporaneous with
the gradual recognition by the public of the futility

of alchemistic aims, and of the deceptions practised by the pseudo-alchemists. Yet there still exist a few individuals who, rejecting the results and opinions of others, and doing their own thinking, entertain firm beliefs in such chimeras as the perpetual motion, the quadrature of the circle,[1] and the philosopher's stone.[2]

Boyle maintains that it is impossible to state offhand the number of the elements. He further postulates a corpuscular structure of matter, and conceives of chemical combination as an approximation of corpuscles of different natures; of decomposition as the result of the presence of a third kind of corpuscle capable of exerting on one of the combined corpuscles a greater attraction than that exerted by the corpuscles already combined with it. Heretofore the ideas entertained respecting chemical combination had in general been very fanciful. Although the conclusion was not in any way implicated in the Aristotelian conception of elementary principles which held sway, yet the formation of a new substance had been regarded as a special creation following on a real destruction of the combining bodies.

Boyle was the first to distinctly state that the peculiar properties of substances disappear on the occurrence of chemical combination—or rather become merged into the new and unforeseen properties of the

---

[1] "Quadrature of the circle found. Lessons given. Fee five shillings." Extract from advertisement appearing in *Nature*, August 17, 1893.

[2] A good picture of this type of man is given in Balzac's *L'Alchimiste*. See also Dumas, *Memoirs of a Physician*.

compounds formed. In this connection he drew a
hard and fast line between mixtures and true chemical
combinations—a distinction which will be entered into
in more detail in Chapter III.

Boyle further advocated the pursuance of the science
of chemistry independently of utilitarian ends. He
combated the view that it was the handmaid of this
or that science, it was, according to him, a self-con-
tained and independent part of the great study of
nature. In the mere advancement of knowledge for
its own sake he found a sufficient spur to a devoted
study of chemistry.

In the next chapter we shall see how Boyle's ideals
were realised; how the phenomenon of combustion
was disinterestedly studied for the sake of the light
which it was hoped would by this means be shed on
chemical theory generally; and how the adoption and
elaboration of Boyle's conception of the elements was
one of the chief factors in Lavoisier's happy co-ordination of the apparently isolated and vaguely expressed
items of knowledge of his day into a harmonious and
perspicuous whole.[1]

---

[1] The reader, who may be incited by the short sketch of the history
of Alchemy given in the text to a deeper study of the subject, is
referred to the following literature :—

Article "Alchemy," *Enc. Brit.*; Rodwell, "Birth of Chemistry;"
Meyer, "History of Chemistry;" Thomson, "History of Chemistry;"
Kopp, "Geschichte der Chemie," and "Die Alchemie in alterer und
neurerer Zeit;" Berthelot, "Les origines de L'Alchimie;" Ben
Jonson, "The Alchemist." (Morley's Universal Library.)

CHAPTER II.

THE PHLOGISTIC PERIOD AND THE BEGINNINGS OF CHEMICAL THEORY.

THE phenomenon of fire is such an important factor in chemical change, both as an agent and as a result, that the chemists of the seventeenth and eighteenth centuries regarded it as the essential phenomenon of chemistry. To found a consistent and competent theory of combustion was in their opinion almost tantamount to the founding of a satisfactory theory of general chemistry; and in the light of modern science, which teaches that so many important chemical reactions are either changes of the nature of a combustion (*i.e.*, are oxidations) or changes of the nature of the reversal of combustion (*i.e.*, reductions), this opinion finds some justification.

The first theory of chemistry instituted by Becher (1635–1682), and developed by Stahl (1660–1734), was essentially a theory of combustion, and is nothing else than a special development in one direction of Becher's modification of the alchemistic views on the nature of matter.

According to Becher the fundamental constituents of inorganic matter were not sulphur, mercury, and salt, but the three earths, the mercurial or fluid earth, the vitreous or fusible earth, and the combustible or fatty earth.   The latter was called *terra pinguis*.

Stahl especially elaborated this idea of a *terra pinguis*.   According to him all combustible bodies (sulphur, phosphorus, carbon, metals, &c.) were compounds containing as an essential ingredient a fiery principle which he renamed "phlogiston."   In the process of burning this phlogiston escaped from combination often in such quantities and with such intensity as to produce the phenomenon of flame.[1]   In the case of the combustible metals, the residue or earthy powder remaining after the phlogiston had escaped was termed a calx; hence such metals were regarded as compounds of calx with phlogiston.   In the case of what are now generally but loosely known as the non-metals, acids resulted from combustion, hence these were regarded as compounds of acids and phlogiston. It was not at first noticed that the calx or acid, as the case might be, weighed more than the substances from which they had been produced—chemistry was as yet in its purely qualitative stage.[2]

When the calx or acid was reheated with a body rich in phlogiston, such as charcoal, it combined with

---

[1] Death by "spontaneous combustion," which was firmly believed in during the eighteenth century, and has even figured in the fiction of the nineteenth (see *Bleak House*, Preface and chapter xxxii.), met with a very simple "explanation" in terms of the phlogiston theory.

[2] Boyle seems to have been the first to notice this increase in weight. He ascribed it to the combination of the burning body with "igneous particles."

the phlogiston of the charcoal, reforming the burnt substance. Such was the phlogistonist's description of reduction. But other bodies rich in phlogiston, such as sulphur, flour, sugar, effected the transformation of a calx or acid into identically the same metal or non-metal as did charcoal.[1] Hence it was argued that there was but one kind of phlogiston.

This phlogiston was generally regarded as a definite material substance. Stahl looked forward to its isolation, and seems to have expected a solid earthy body insoluble in water. In fact the matter of solubility appeared to be merely a question of a greater or less amount of combined phlogiston. Phosphorus and sulphur, bodies rich in phlogiston, are insoluble in water, but the acids produced from these substances by the escape of phlogiston are eminently soluble. The quantity of phlogiston in a substance was in fact believed to condition not only its solubility but all its properties—its activity or inertness, its stability or instability, its acidity or basicity, &c.

Although Stahl seems to have been in favour of a solid phlogiston, others regarded it as gaseous, and indeed went so far as to identify it with hydrogen. When metals dissolve in acids a gas escapes, which if collected and heated with the calx recoverable from the solution—the calx in this instance really being

---

[1] In very early times it was known that wheat has the power of revivifying a metal from its calx or ashes, and it is said that it was partly on account of this property that wheat was made the popular symbol of the resurrection.

a salt—restores the metal.   But metal is calx and phlogiston; the gas disengaged on dissolving metals in acids is therefore phlogiston.

The doctrine of phlogiston as just sketched, after reigning triumphantly for some half century, finally succumbed to the increasing knowledge of the constitution and nature of the atmosphere. This knowledge shed quite a new and certain light on the phenomena of combustion, and was co-ordinated by Lavoisier into the theory of combustion held at the present day.

Let us now trace the chief stages in the growth of our knowledge of the atmosphere.

Robert Hooke (1635 – 1703), a contemporary of Becher, and sometime assistant to Boyle, published in 1665 his conception of combustion in his *Micrographia*. He believed that there existed in air a fractional quantity of the same kind of gas as is obtained by heating saltpetre, or nitre, and that combustion consisted in the solution of the combustible body by this gas. Hooke pointed out many analogies between combustion and the solution of solids in liquid menstrua.[1]

Mayow (1645–1679), an Oxford physician, worked out in more practical detail the ideas of his contemporary Hooke. He showed that when a metal burns in air, the volume of air is actually lessened, and that

---

[1] It is interesting to note that Hooke was on the point of anticipating Rumford. He clearly stated that fire or flame is not an element, but a phenomenon resulting from the agitation of particles.

the calx residue weighs more than the original metal.
This he explained by suggesting that the metal had
combined with the particles of the nitre air, as he
called it, present in common air, leaving a residue of
inactive air.   The same fact was glossed over by the
phlogistonists, in terms of the assertion that phlogiston
was a principle of levity, and that therefore its escape
from a body during combustion rendered the body
heavier.   And so supremely popular was the phlogiston theory at the time, that the reasonableness and
common sense of Mayow's explanation had no weight
against the transcendental artificiality and uniqueness
of its own.

Mayow's experiments were so much to the point,
that it is hard to see how it came about that their full
significance was not recognised till they were practically rediscovered piecemeal a century afterwards.

The next advance towards an accurate knowledge of
combustion was Black's discovery and investigation of
what he called "fixed air"—the gas now known as
carbonic acid gas, or carbonic anhydride.

The attention of the medical world was about this
time (1728–1799) directed to quicklime, magnesia, and
allied substances, in the rôle of efficacious remedies in
the treatment of stone in the bladder, and Black, being
greatly interested in medicine, undertook a chemical
investigation of these new remedies.   Up to this time
the carbonates of lime, magnesia, and the alkalis had
been regarded as elements, or rather simple bodies,
which when heated formed quicklime, caustic alkali,

&c., in virtue of their combination with the phlogiston escaping from the burning coal.

Black chiefly investigated what was then called mild magnesia, *i.e.*, magnesium carbonate. He first proved that when the carbonate is heated, it loses weight. This loss is not due to a combination with phlogiston of negative specific gravity, but to the expulsion of a gas which he called fixed air, because it could be again refixed in a sort of latent condition by the resulting caustic magnesia. He showed that the same gas is produced when chalk is calcined or treated with acids, that it exists in the breath of animals, and is evolved in large quantities during fermentation. Further, that animals die when placed in it, and that it does not support combustion. Nothing regarding the composition of this gas was, however, hinted at.[1]

The next step towards the full light was taken by the English divine Priestley (1733–1804). Priestley did not allow his Calvinistic doctrines to usurp his whole interest and leisure, but he gave much attention to chemistry, making a speciality of that branch of it dealing with the different kinds of air—or as we should say now, the different kinds of gases.

While taking charge of a chapel in Leeds, he happened to live near a brewery, in which he spent much time studying the properties of fixed air, which Black had shown to be produced in large quantities during fermentation. On being compelled by circumstances

---

[1] Black's paper has been republished in convenient pamphlet form by the Alembic Club (Sampson & Co.).

to move from the purlieu of this laboratory, he did not drop his interest in fixed air, but began to make it artificially from chalk as taught by Black. In order to store and examine gases, he made extended use of and popularised Mayow's method of collecting them over water in what he called a pneumatic trough, a method still in vogue. He soon noticed that the fixed air dissolved to a considerable extent in water, and that the solution had medicinal properties. From this discovery dates the manufacture of artificial mineral waters.

Priestley, having fairly exhaustively worked out the properties of fixed air, turned his attention to new airs. In this investigation he seems to have been guided by no principle of selection, but at haphazard he subjected to the action of the sun's rays concentrated by a lens any compound that he happened to think of or chance upon; the gas, if any, produced being collected over mercury.[1] One of the substances thus experimented on was the so-called red precipitate. From this, in 1772, he obtained a gas in which inflammable bodies burned vigorously, and "which had all the properties of common air, only in much greater perfection." He gave it the name of dephlogisticated air, and regarded it merely as common air quite free of impurities and admixtures. He did not regard it as a constituent of air as Hooke and Mayow some years previously had done with truer insight. Air was to Priestley essentially a simple sub-

---

[1] A statue of Priestley in Birmingham represents him conducting this operation.

stance, and his dephlogisticated air was merely common air in a state of great purity. In 1774, being in Paris, he demonstrated to Lavoisier his method of making dephlogisticated air. Some years later, as we shall see, Lavoisier made great use of this knowledge, and re-christened dephlogisticated air, giving it its present name, oxygen.

Priestley was a confirmed phlogistonist. The following is something like his idea of what takes place during combustion. When a substance burns, phlogiston escapes from it into the air, which is invested with a great affinity or longing for this principle of fire. Substances cannot burn out of contact with the air, or other substances, such as nitre, which possess this strong attraction for phlogiston. · But in proportion as the air round a burning body becomes more and more charged with phlogiston, the poorer a supporter of combustion does it become, till finally it may become so phlogisticated—so noxious, as he expressed it—that it will actually quench a flame immersed in it. The air obtained from red precipitate was virgin air untainted by any phlogiston, and so energetically did it long for this principle of fire that even at ordinary temperatures it drew it away from several substances, thus causing them to tarnish. Iron nails, as we know, soon tarnish, i.e., rust, when immersed in impure oxygen.

If carbon is heated with metallic calx it gives up the phlogiston in which it is rich to the calx, reforming the pure metal from it. But according to Priestley the air that has surrounded a burning body is rich in phlogiston ; it then, by a parity of reasoning, ought to reduce metals from their calces. If Priestley had but tried the experiment here suggested, he would have met with such a positive failure that one can hardly conceive how he could have continued to cherish his views on the nature of combustion.[1]   It will be noticed that Priestley did not attempt to explain why the air diminished when combustion took place in a confined space.

Priestley, essentially a *preparateur*, also experimented in his own casual way on nitrous air (nitric oxide), vitriolic acid air (sulphur dioxide), muriatic acid air (hydrochloric acid), alkaline air (ammonia), and inflammable air (hydrogen), which latter had been discovered by Cavendish in 1766 and·recognised by him as an individual gas.

In his experiments on inflammable air, Priestley very nearly anticipated Cavendish's discovery made three or four years afterwards, of the composition of water. Indeed, Priestley's mind was so saturated with the importance of the chimera phlogiston,[2] that for once at least in his life he failed of a customary good fortune to which he alludes in the following words :— "In looking for one thing I have generally found another, and sometimes a thing of much greater importance than that which I was in quest of."

---

[1] It is easy to point out many similar inconsistencies in the theory of phlogiston. For instance, the ash from burnt charcoal ought to weigh more than the original charcoal.

[2] Priestley's last work is entitled *The Theory of Phlogiston Established.*

In experimenting with the various airs, Priestley always tried their effects on animal and vegetable life. For this purpose mice and sprigs of mint seem chiefly to have been victimised. In this connection it is interesting to recall that he distinctly noted that wonderful piece of natural economy whereby the gaseous waste-products of animals serve as the food of plants, and the atmosphere is maintained in a state of freshness and purity compatible with the prolonged existence of life.

It is a great source of regret to read that Priestley, who by his blind and undirected enthusiasm for discovery laid some of the foundation-stones of modern chemistry, was so little appreciated by a large section of his countrymen that he finally left his native land with the remark, "When the time for reflection shall come, my countrymen will, I am confident, do me more justice."

Before we pass on to the brilliant co-ordination of the facts of combustion and their true explanation by Lavoisier, we must mention briefly the work of that peculiarly ascetic philosopher, the Hon. Henry Cavendish—work which furnished Lavoisier with some of his most weighty data.

In 1766 Cavendish discovered that when about two volumes of inflammable air are exploded with one volume of dephlogisticated air, water is produced. Further, if ordinary atmospheric air is used instead of dephlogisticated air, water is still produced, but a residue of phlogisticated air, *i.e.*, what we now call nitrogen, always remains. Hence Cavendish argued that air was not a simple individual substance, but a mixture of phlogisticated and dephlogisticated airs. It would seem, however, that Cavendish did not clearly grasp what seems to us such an evident conclusion from his experiment, viz., that water is a compound of inflammable and dephlogisticated airs.[1] He had been brought up on the phlogiston doctrine, and could never thoroughly free himself from its trammels. Although he saw good reasons for supposing that when a metal burns in the air it combines with the dephlogisticated air to form a calx leaving the phlogisticated air, yet he could not find it in his heart to give support to such a heresy. He therefore concluded with Priestley that the metal phlogisticated the residual air, but slightly differed from that scientist in admitting that *some* of the air in a dephlogisticated state combines with the metal. Combustion was a case of a partial combination associated with and attended by a contamination; not of a total combination attended by an elimination.[2]

It is remarkable that the most fearful reign of social anarchy and bloodshed the modern world has known should have bequeathed to us through Lavoisier so unexpected a gift as a remodelled and thoroughly scientific theory of chemistry. Lavoisier was born in Paris in 1743. His peaceful life of scientific research, passing amidst all the horrors culminating in the reign of terror, was finally ended on the guillotine.

[1] James Watt seems to have been the first to arrive at this conclusion.

[2] The Alembic Club has reprinted in convenient pamphlet form some of the more important of Cavendish's papers.

About the year 1770 Lavoisier turned his attention to combustion. His great success in this field of study must be largely attributed to the importance he placed on the incessant use of the balance as an instrument of research ; an instrument with which he, early in his scientific career, established the principle of the conservation of mass [1]—a principle lying at the foundation of the science of to-day, and familiar to all. *In no chemical change is matter either created or destroyed; the sum of the masses of the factors of a chemical change being identically equal to the sum of the masses of the products. The form of matter can be changed, but not its quantity.* This principle, which strikes the materialistic mind of to-day so much in the character of a truism, involved quite a revolution in the settled views of the end of the eighteenth century. As has before been stated, the production of new forms of matter was regarded as an act of special creation, having no quantitative relation to the destruction or annihilation which necessarily preceded it. In Lavoisier's time, too, heat or caloric was popularly regarded as material, and consequently its greater or less *rôle* in any chemical change would presumably more or less affect the mass relations. But Lavoisier proved that the mass of a compound AB is, within the limits of experimental error, exactly equal to the mass of A and the mass of B combining to form that compound, and is quite independent of the heat evolved or absorbed during the combination.

[1] This principle was assumed in many of the ancient philosophies.

Although Black and Cavendish had done accurate pieces of quantitative work, yet it is with Lavoisier that chemistry essentially passed from the qualitative to the quantitative stage.   It was he who ever insisted on the paramount importance of the question, "How much?" in every chemical inquiry.

In connection with combustion, Lavoisier first confirmed the fact that when a metal burns it increases in weight, and then proceeded to prove in the case of tin that the increase in weight is due to air absorbed, and is equal to the weight of the absorbed air.   He noticed that it was not the air as a whole that was absorbed, but only a constituent thereof, the unabsorbed constituent being different from common or fixed airs. From classical experiments on red precipitate (HgO), suggested by Priestley, he concluded that the constituent absorbed by burning bodies was nothing else than Priestley's dephlogisticated air.

While this air can be expelled from the red calx of mercury by mere heating—a point whose explanation had always baffled the phlogistonists—the calces of other metals had to be heated with carbon, and the air disengaged under these conditions Lavoisier proved to be identical with Black's fixed air, which therefore must consist of carbon and dephlogisticated air, and which he therefore renamed carbonic acid gas.   The carbon combined with the dephlogisticated air of the calx, just as it did with that of the atmosphere when burned therein. This same carbonic acid gas Lavoisier showed to be formed during the respiration of animals, thus transforming a suspicion which had arisen in the mind of Paracelsus into a certainty. Mayow a century previously had come to the same conclusion, though not in so definite a manner.

Since the combination of dephlogisticated air with the majority of substances investigated by Lavoisier produced acids, he renamed the gas oxygen, or the acid producer; a nomenclature which modern research has shown to be not altogether appropriate.

On burning phosphorus in a confined volume of air, Lavoisier found that after about $\frac{1}{5}$ of the air had disappeared the phosphorus was extinguished, and all other combustible substances refused to burn in the residual gas. He regarded this noxious air as a true constituent of common air actually present therein before the burning, and called it first "moufette atmospherique," and then later "azote," because animals died when immersed in it. Rutherford in Edinburgh simultaneously isolated this gas by treating with caustic potash solution air in which animals had been confined for some time. He gave it the more popular name nitrogen. From his experiment with burning phosphorus, Lavoisier rightly concluded that air was a mixture of about $\frac{1}{5}$ oxygen and $\frac{4}{5}$ nitrogen or azote.

Having given a simple explanation of calcination or oxidation and reduction by carbon, Lavoisier now turned his attention to the process of reduction, $i.e.$, the reforming of the metals from their calces by means of hydrogen. To this end he repeated Cavendish's experiment on the composition of water, rightly interpreting the results and completing the proof by passing steam over heated iron, thus obtaining a calx and inflammable air which he thenceforward called hydrogen. When a calx is heated with hydrogen, the metal reappears, not because the calx takes up the phlogiston, in which the hydrogen is rich, but because the hydrogen wrests the oxygen from the calx and forms therewith invisible steam.

Lavoisier regarded the solution of a metal in a dilute acid as a reaction taking place in two stages. First, the metal decomposed the water, setting free the hydrogen, and combining with the oxygen to form a calx; and then this calx combined with the acid, forming a salt soluble in the water holding the acid in solution.

Here truly was a complete *renversement* of current ideas. To the phlogistonist the metal or combustible was more complex than the calx or product to which it gave rise. The metal or combustible consisted of matter plus phlogiston, the calx or product was metal or combustible minus phlogiston. To the Lavoisierian or antiphlogistic school, however, the calx was more complex than the metal—it was a compound, while the metal was an element.

When we look back on Lavoisier's life-work, we see it to consist of co-ordination and generalisation rather than discovery. He made use of the facts brought him by what Huxley has called the hod-carriers of science. In making use of these contributions he sometimes, unfortunately for his moral reputation, forgot their sources, and claimed them as original. Lavoisier's attitude towards Priestley, Cavendish, and others, whose data he used, is still considered a subject for severe criticism.

Undoubtedly a great factor in Lavoisier's success was his unconditional rejection of the old, vague, elementary principles of the alchemistic era, and his adoption of the conception of elements and compounds as developed by Boyle, and now generally received.

The theory of phlogiston was doomed. Lavoisier's explanations steadily gained in favour, and the year 1785 marked the complete ascendency of his antiphlogistic views. In France his teaching was received partly from intellectual conviction, partly from a feeling of patriotism, stimulated by the title *La Chemie Française*, which Fourcroy was presumptuously pleased to bestow on his countryman's theory.

The majority of British and German chemists also expressed their allegiance to the new theory before its illustrious founder prematurely passed away.

CHAPTER III.

CHEMICAL CLASSIFICATION. MIXTURES, COMPOUNDS, ELEMENTS.

THE fundamental classification of chemistry is comprised in the two expressions, homogeneous or pure bodies, and non-homogeneous bodies or mixtures. Everything material necessarily falls into one or other of these two classes.

It is sometimes quite erroneously stated that only the homogeneous substances belong to the domain of chemistry. This statement probably originates in the fact that all the fundamental chemical laws have reference to, and are only true for, homogeneous bodies; but homogeneity is the exception, not the rule in nature,[1] and chemistry is a natural science. Matter invariably comes into the laboratory in a state of non-homogeneity or mixture, and it is always the first and often the most difficult task of the chemist to resolve such a mixture into its homogeneous constituents.

---

[1] Witness the derivation of the word metal. According to Pliny this word is derived from μετ' ἄλλα signifying *together with other things*. There is, however, another possible derivation of the word metal, viz., from μέταλλον = *a mine,* connected with μεταλλάω = *I search for diligently.*

The study of coal tar, for example, has resulted in most important advances in the theory of chemistry. Yet the study of coal tar *as a whole* would have been quite barren of valuable results. Tar being a variable mixture, the investigation of its properties would have led to inconstant results, and made impossible any attempts at productive generalisation. But by the process of fractional distillation (*vide infra*), coal tar has been sorted into a number of homogeneous constituents, and it is the study of these constituents individually, not of tar as a whole, that has profitably engaged the attention of chemists.[1] Thus benzene

[1] The following scheme represents the approximate percentages obtained, and the stages passed through in the separation (see *post*, p. 47) of six of the more important homogeneous constituents of coal tar :—

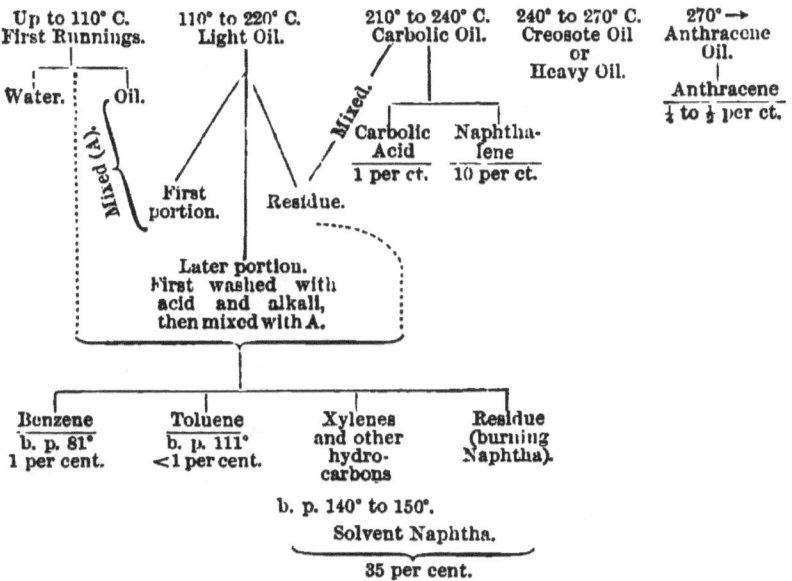

is a definite substance with fixed and invariable properties, and its constants determined for a preparation from any one coal tar will hold good for a specimen obtained from any other tar.

What, we may now inquire, is the distinguishing feature of the class of non-homogeneous bodies? Briefly stated it is this; a non-homogeneous substance or mixture can be resolved into unlike portions by processes of mere sorting, and the sum of the energies of the sorted portions is equal to the energy of the mixture from which they were sorted.[1]

If one simply mixes a red powder A with a green powder B (no heat evolution or absorption accompanying the process) the mixture can be resolved into its constituents again by patient "hand sorting" under the microscope.[2] If flour and sugar be mixed, they can be separated again by treating the mixture with water, in which the sugar alone is soluble. The mere mixture of flour and sugar does not alter the properties peculiar to each of these substances in the pure state.

Even when all the constituents of a mixture are soluble in a given medium, they may often be com-

---

[1] Mixtures or emulsions of such liquids as are ordinarily spoken of as *immiscible*, *e.g.*, oil and water, are only temporary. Such mixtures possess more energy than the separated constituents; but this surplus of energy, depending on the area of the surfaces of separation of the different liquids, sooner or later runs down into heat, and contemporaneously the mingled liquids sort themselves into easily separable layers. [See Maxwell, *Theory of Heat*, chap. xx.]

[2] Dextro- and laevo-tartaric acids were first separated from each other and obtained pure by this method.

pletely separated by repeated crystallisation from that medium. Thus when an aqueous solution of potassium chloride and potassium chlorate is evaporated, the less soluble chlorate crystallises out first, and afterwards the more soluble chloride. This method of separation is actually adopted in the commercial preparation of chlorate of potassium. At Stassfürt a mineral composed of potassium and sodium chlorides is found in quantity. In order to separate the constituents, a cold saturated solution is first made of the mixture. This solution is then raised to the boiling point. In proportion as the water evaporates the sodium chloride crystallises out. After a certain time the solution is allowed to cool, when potassium chloride separates out. For this salt, while much more soluble at high temperatures than common salt, decreases in solubility very rapidly indeed as the temperature is lowered. By alternately repeating the boiling and cooling, large quantities of the chlorides are separated.

Again, oxygen and nitrogen are both soluble in water, but oxygen to a greater degree than nitrogen. Common air is a mixture of these two gases, consequently the " air " dissolved by water (which " air " is disengaged again on boiling the water) is much richer in oxygen than common air. If the process of treating water with air under pressure, and then boiling the resulting solution, were repeated often enough fairly pure oxygen could be eventually sorted out. An investigation of the results of a single operation of the above process will be instructive and will afford an opportunity of illustrating important laws, one discovered by Dalton, and the other by Henry and Dalton. The data requisite for this investigation are as follows : —

		Per cent.
Composition of air by volume, oxygen . .	20·9
,, ,, ,, nitrogen . .	79·1
Approximate coefficient of absorption of oxygen .	·04
,, ,, ,, nitrogen	·02

The coefficient of absorption of a gas denotes that volume thereof (measured at 0° C. and under a pressure of 760 mm. of mercury) which is dissolved by one volume of water when an excess of the gas is presented to the water under a pressure of 760 mm. of mercury. Thus one cubic foot of water dissolves ·04 of a cubic foot of oxygen (measured at 0° C. and 760 mm.) when an excess of the gas is allowed to stand over the water at normal pressure.

Dalton's approximative law [1] of partial pressures (1802) states that the total pressure of a mixture of gases is the sum of the partial pressures exerted by the constituents of the mixture in the space occupied by the mixture. Suppose the air exerts a pressure of 760 mm. Then, from Boyle's law and the above data, the partial pressure of the oxygen P_o must be given by the equation

$$P_o \times 100 = 20\cdot9 \times 760.$$
$$P_o = 760 \times \cdot209 \text{ mm.}$$

Similarly the partial pressure of the nitrogen P_n is

[1] "An approximative law expresses only a portion of a complex phenomenon—the limit towards which the phenomenon aims."

equal to $760 \times \cdot 791$ mm. In conformity with Dalton's law, it will be noticed that $P_o + P_n = 760$ mm.

Henry and Dalton's law (1803–1807) is an approximative law which states that the masses of the less soluble gases dissolved at ordinary temperatures are proportional to the partial pressures exerted by these gases.[1] Let d be the density of oxygen at $0°$ and 760 mm. Then one cubic centimetre of water dissolves $\cdot 04d$ grams of oxygen when the pressure is 760 mm. By Henry and Dalton's law it will therefore dissolve

$$\frac{\cdot 04d \times 760 \times \cdot 209}{760} \text{ or } \cdot 04d \times \cdot 209 = \cdot 00836 \times d \text{ grams,}$$

when the pressure is $760 \times \cdot 209$ mm. The volume of this oxygen is $\cdot 00836$ cc. Similarly $\cdot 02 \times \cdot 791$ cc. of nitrogen are absorbed by one cubic centimetre of water. Hence the composition of the dissolved air is approximately

Oxygen	.	.	.	. 34·5 per cent.
Nitrogen	.	.	.	. 65·5 per cent.

In other words, whereas air contains only about 21 per cent. of oxygen, the dissolved air contains about $34\frac{1}{2}$ per cent.[2]

---

[1] It is frequently stated that the law of Henry and Dalton furnishes a criterion as to whether a gas is physically or chemically dissolved. If the law is obeyed, then it is said to be a case of physical absorption, otherwise it is chemical change. Such a hard and fast distinction cannot be maintained. Solution is a complex phenomenon, and even in the simplest cases it seems almost certain that true chemical action and simple physical change are inextricably involved (see p. 55). Further, it is probable that at low enough temperatures no gas would be found to obey Henry and Dalton's law.

[2] If the process indicated in the text were to be repeated with this

Many mixtures may be resolved by taking advantage of differences in the volatilities of their constituents. When in the analysis of an inorganic substance by the usual methods we come to test for potassium, it is necessary before performing the test to remove the salts of ammonium which have been added earlier in the course of analysis. This is effected by evaporating the solution to dryness and then heating the residue to low redness, when the more volatile ammoniacal salts will pass off, leaving behind the less volatile potassium salts.

A mixture of liquids can often be sorted by fractional distillation; that is, by taking advantage of differences in the boiling point of its constituents. If a mixture of ether (B.P. 35°) and aniline (B.P. 181°) be heated in a suitable apparatus, the temperature of the mixture rises quickly till the neighbourhood of the boiling point of ether is reached, when it remains fairly constant till nearly all the ether is distilled off. The temperature then rises rapidly to the boiling point of aniline, when nearly pure aniline is obtained as a distillate. It is not in general possible to effect in this way a complete separation of

air of $34\frac{1}{2}$ per cent. oxygen, the composition of the air resulting from a second solution in water would be approximately

47·5 Oxygen,
52·5 Nitrogen,

and a third solution in water would make the composition of the dissolved air approximately

75 Oxygen,
25 Nitrogen.

mixed liquids in one operation, it is only when the process has been repeated systematically several times that perfect separation is attained.

This process of fractional distillation is the first and most important stage in the resolution of coal tar into its homogeneous constituents (see p. 41); the final isolation of these constituents in a state of complete purity being effected by other sorting processes adapted to the individual cases.

As will readily be conceived, a mixture of liquids possessing the same or nearly the same boiling points does not yield to this method of fractional distillation, and indeed in many cases where the boiling points of the constituents of a mixture differ fairly widely, separation is found to be impracticable. Let A and B be the two constituents of a liquid mixture; then it would appear that a separation of A and B by fractional distillation is only possible when the boiling point of every conceivable mixture of A and B lies intermediate between the boiling points of the free constituents A and B. This is the case with ether and aniline, and therefore all mixtures of these two substances can, as already described, be separated by fractional distillation. But it is not the case, for instance, with ethyl alcohol (B.P. 78°) and water; hence mixtures of alcohol and water are not completely resolvable by fractional distillation in spite of the large difference between the boiling points of the liquids.[1]

---

[1] A mixture of 97 per cent. ethyl alcohol with 3 per cent. water has a higher boiling point than either pure alcohol or pure water.

Many mixtures may be resolved by taking advantage of differences in the densities of their constituents. This principle finds application in the "panning out" of gold from its ores, in the sorting of diamonds from their specifically lighter matrix in the so-called pulsators, and in the resolution of various mixtures by centrifugalisation.[1]   In the analysis of a soil the clay is separated from the sand by stirring up the soil in water. The sand sinks to the bottom of the containing vessel, while the clay remains suspended in the supernatant liquid, which is poured off and filtered. Such methods of resolution, wherein an inert menstruum is called into play, are often grouped together under the title of elutriation methods.[2]

A mixture of gases of different densities can also be partially resolved by a process introduced by Graham (1805–1869), and called by him atmolysis—

Hence pure anhydrous alcohol cannot be obtained by fractional distillation. Suppose we start with a mixture of 50 parts water and 50 parts alcohol. Fractional distillation will resolve this into two portions : (1) 48·45 parts of pure water, and (2) 51·55 parts of 97 per cent. ethyl alcohol (50 parts alcohol, and 1·55 parts water). In other words, it is fairly easy to get in a pure state some of the water mixed with the alcohol, but it is impossible to get alcohol in an anhydrous state from an aqueous mixture. To obtain pure anhydrous alcohol recourse must be had to distillation of the 97 per cent. alcohol over anhydrous baryta.

[1] The centripetal force necessary to keep a particle of mass m moving in a circular path of radius r with angular velocity a, is numerically equal to the product mra^2. Consequently the more massive, or rather, the denser particles of a mixture will have a greater tendency to fly tangentially off a revolving plate than the less dense ones.

[2] For very neat applications of elutriation methods to the proximate analysis of minerals see *Nature*, xliii. p. 404, and xlix. p. 211.

a rather misleading nomenclature in this connection. The lighter a gas is, the more quickly does it pass through a porous partition, *cæteris paribus*.   In the annexed diagram *cc'* represents a glass tube through the axis of which runs a porous tube (*e.g.*, the stem of a churchwarden pipe) TT'.   The side tube *c'* of the glass is in connection with an exhausting pump.   If electrolytic gas, which is a mixture of hydrogen and oxygen in the proportion of two volumes of the former to one of the latter, be passed slowly through the tube

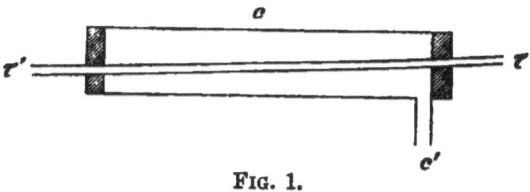

FIG. 1.

TT', the pump will extract from *c* a mixture richer in the lighter gas hydrogen than the electrolytic gas used, while the mixture issuing from TT' will be richer in oxygen.   Finally a limit will be reached whereat the repetition of the process will be unproductive of further resolution. · The exact statement of the connection between the densities of gases and their rates of passage through a thin septum of graphite, or better still, biscuit ware, is known as Graham's law of diffusion.   The rates of diffusion of gases under the same pressure vary inversely as the square roots of their densities; or algebraically stated—

$$\frac{\text{rate of diffusion of gas } x}{\text{rate of diffusion of gas } y} = \frac{\sqrt{\text{density of } y}}{\sqrt{\text{density of } x}}$$

D

This law is not actually realised in practice; it would be absolutely true for "perfect gases" (*i.e.*, ideal gases which obey Boyle's law perfectly) diffusing through an infinitely thin septum.[1]

We may also separate some mixtures by taking advantage of their different diffusibilities, as in the process of dialysis, also invented by Graham. When substances are in solution, the rates at which they pass through animal or vegetable membranes[2] differ widely. Substances which can crystallise pass through such membranes readily.[3] On the other hand, solutions of bodies which do not crystallise but exhibit conchoidal fracture, are either quite incapable of passing, or only pass with the greatest difficulty, through such membranes. The former readily diffusible class of bodies Graham called

---

[1] This law, in common with the other gaseous laws of Boyle, Charles, and Avogadro (see p. 89), is a necessary consequence of the kinetic theory of gases. This theory simply postulates that the pressure of a gas is due to the impact on the walls of the containing vessel of the small similar particles of which the bulk of the gas is made up. Reasoning thence, the expression $\sqrt{\dfrac{3p\mathrm{D}g}{d}}$ is arrived at by pure mathematics for the average velocity of progressional movement of the particles of a gas whose density is d, under a pressure of p cm of mercury of density D, at a place where the acceleration due to gravitation is g. Since the rate of diffusion of a gas varies directly as the average velocity of movement of its particles, the law of Graham at once follows.

[2] For purposes of dialysis, a parchment made by soaking unsized paper in sixty per cent. sulphuric acid for a short time, is in general use.

[3] Striking exceptions to this generalisation are furnished by hæmoglobin, the colouring matter of the blood, and vitellin, an albumen occurring largely in plants. Both of these substances, though crystalline, are non-diffusible. It has also been stated that egg albumen—a colloid almost as typical as glue—can be obtained in the crystalline state.

crystalloids; to the latter non-diffusible class he gave the name colloids, from the Greek word (κόλλα) for glue, their typical representative. As an example of the application of the method of dialysis we will take the preparation of the important drug "liquor ferri dialysati." Like many colloid bodies, ferric hydroxide can exist in a soluble or hydrosol form, and in an insoluble or hydrogel form. A solution of the hydrosol form constitutes the drug above mentioned, and it is prepared as follows. The hydrogel or ordinary form of ferric hydroxide is soluble in a solution of the crystalloid ferric chloride. Such a solution is prepared and then placed in a dialyser (Fig. 2). This consists of a U tube uu, made out of parchment paper suspended from a rod rr in a large vessel contain-

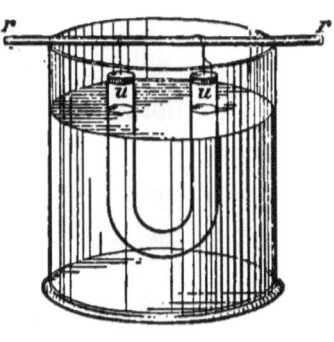

FIG. 2.

ing water. The crystalloidal ferric chloride diffuses into the surrounding water, which should be frequently changed; while the hydrosol form of the colloidal ferric hydroxide[1] remains behind in the U tube in a pure state.

This process of dialysis is frequently resorted to in toxicological investigations for the purpose of separating the crystalloidal poisons from the colloidal contents

---

[1] It would appear that liquor ferri dialysati is not a solution of pure ferric hydroxide, but of a hydrosol ferric hydroxychloride.

of the stomach and intestines which would mask the reactions of the poisons.

The method of refrigeration is also largely used in the resolution of mixtures. Thus the impure benzene obtained by fractionally distilling coal-tar is freed from admixed toluene by surrounding it with ice. The benzene solidifies, and can be separated from the still liquid toluene. Similarly pure sulphuric acid is prepared from the aqueous product obtained by evaporation in platinum vessels. Quite recently this method of purification by refrigeration has been applied on a large scale to the purification of organic bodies, such as ether, chloroform, &c., by Pictet and Liebreich.

So far we have dealt with what may loosely be called mechanical methods of resolving mixtures. These can, of course, also be resolved by chemical means, but in this case one or other of the constituents of the mixture necessarily suffers a permanent change. Thus, a zinc compound in solution can be freed from an admixed copper compound by passing sulphuretted hydrogen through the acidulated mixture. The copper compound is thus changed into insoluble copper sulphide, which can be filtered off, leaving the zinc compound in solution.

A number of very subtle chemical methods of resolving mixtures of very closely allied bodies have recently been introduced under the generic title "fractionation." These methods will receive more detailed notice when we speak of the elements.

It will now be clear that in a mixture the constituents

conserve the properties peculiar to themselves in the free state. Hence it follows that a mixture when subjected to certain operations shows in general differential results—part dissolves, part does not; part sinks, part floats—so that it can be sorted into its constituents by taking advantage of sufficiently pronounced differences in degree of a common property of its constituents. In thus studying the resolution of mixtures, we have gained some knowledge of the more general and important methods employed by chemists for purifying substances.

Turning now to homogeneous substances. These, as was first hinted by Boyle, can be divided into the two classes, elements and compounds. The elements are practically the units of chemistry; according to definite laws, they unite in manifold combinations and permutations to form the now almost infinite number of known compounds; but they defy all attempts to resolve them into dissimilar constituents. Mercury, e.g., cannot in any known way be split up into dissimilar portions, each portion weighing less than the mass started with. In short, elements constitute that class of homogeneous substances which is amenable to synthesis only; elements cannot be analysed. They constitute what may metaphorically be called the alphabet of the science.

But if compounds are composed of and are resolvable into elements, how, it may be asked, do we distinguish them from mere mixtures? Put briefly, the distinction is this: mixtures can be resolved by processes of mere sorting; compounds cannot be so resolved. When elements combine chemically, the properties of the resulting compound are only remotely functions of the properties of the elements combining. The altogether new properties of the compound are not by any means the combined properties of the constituents; they show no partial or differential character. A compound is either wholly soluble or wholly insoluble in a given menstruum; it is either wholly magnetic or wholly non-magnetic; under the highest magnifying power attainable it is optically homogeneous. A knowledge of the salient physical properties of the free elements constituting a compound does not assist us in our endeavours to resolve that compound into its constituents, as is the case with mixtures. A mixture can be resolved into its constituents without involving a simultaneous and permanent change in the matter separated or other matter; whereas the analysis of a compound in general demands the permanent change of at least one constituent of the compound as well as of extrinsic matter. Thus with respect to the preparation of copper from its compound copper oxide, the knowledge that copper is a comparatively non-volatile insoluble body, while oxygen is a soluble gas, avails us nothing; and the realisation of the project absolutely demands the simultaneous and permanent change of some such substance as carbon or hydrogen into oxidised products.

Though the distinction between mixture and compound is in theory easily enough grasped, yet it cannot be denied that in practice it is often extremely difficult, if not impossible, to state definitely whether in particular cases we are dealing with the one or the other class of matter. Physiological chemistry furnishes us with several examples of this difficulty. Is hæmoglobin a "chemical unit" (*i.e.*, homogeneous), or is it a mere mixture of the substances globin and hæmatin, which it so readily yields? Is egg albumen — the most thoroughly investigated of the proteids — a true chemical compound, or is it a mixture of several individual but closely allied proteid substances? Are the varying results of its analyses made by different experimenters to be ascribed solely to accompanying difficultly removable impurities, or to the fact that it is essentially a variable mixture? Is gluten—the sticky constituent of dough—a definite substance, or is it a mixture of the proteids pre-existent in flour but changed by ferment action? To such questions physiologists cannot at present vouchsafe definite answers; indeed it may be said that one of the most formidable obstacles to the rapid advance of physiological chemistry is the uncertainty as to whether many of the fundamental substances with which it deals, and which have received definite names, are, as a matter of fact, "chemical units" (*i.e.*, chemically pure bodies).

Again, the question is often asked, is a solution of salt in water really a mere mixture of salt and water particles, or does it contain what may be called low-grade unstable compounds of salt particles and water particles disseminated throughout the general bulk of the solution ? Is solution a purely physical change in aggregation, or does it involve true chemical combinations between the particles of the solvent and those of the dissolved substance ?

The investigation of the nature of solution takes a most prominent place in the chemistry of to-day. While some chemists are attacking the subject from one side by purely physical methods, and summing up their results in terms of partial and provisional theories involving only physical ideas, others are assailing the subject from a purely chemical standpoint. Yet this statement must not be interpreted as asserting the existence of two rival and mutually exclusive schools on the question of solution. All chemists are pretty well agreed that solution is neither a simple physical nor a simple chemical change, but that it is a very complex process involving simultaneous physical and chemical changes. The exact point at issue is the actual state of dissolved substances, *i.e.*, *the constitution of solutions.*

The only satisfactory explanation that has as yet been given of the results of investigations of solutions from their physical aspect, makes the solvent a passive medium in which the discrete molecules of the dissolved substance (or the products of dissociation of these molecules, the ions) uniformly distribute themselves. From this point of view the solvent bears much the same relation to the dissolved substance as the space defined by the walls of a vessel does to the gas it contains ; solution, in fact, shows an analogy to the evaporation of a liquid in a confined space.

On the other hand, investigations carried out from a chemical standpoint seem to demand the recognition of a definite chemical interaction between, as distinguished from a mere interpenetration of, the molecules of the dissolved substance and those of the solvent. When a salt is dissolved in water it appears as if the salt molecules combine with the water molecules to form definite yet very unstable chemical compounds—true liquid hydrates.[1]

So far then it is clear that the two lines of investigation lead to different views regarding the structure of solutions and the condition of the dissolved substance. But it should be remembered that the more or less mechanical conception of solution has arisen from the exclusive study of very dilute solutions, while the investigations leading to a more chemical *rationale* have been chiefly carried out with strong solutions.

But little doubt can be entertained that future research in this field will either establish a continuity and relationship between these confessedly partial and apparently diverse hypotheses, or by so modifying one or both, will fuse them into a consistent and comprehensive theory of one of the most common but least understood of physico-chemical phenomena —solution.

Twilight is distinct from daylight, but one would

---

[1] The constitution of solutions according to this view will be more fully developed in the chapter on chemical equilibrium.

hardly undertake to say where the one begins and the other ends.   So is it with the distinction between mixtures and compounds.   The extremes of each class contrast strongly enough; but through the intermediacy of what are called molecular compounds [1]— a convenient term with as yet a very vague connotation—the one class merges imperceptibly into, and blends with, the other without break of continuity. [2]

The following rough analogy may serve to fix in the mind the fundamental differences between mixtures on the one hand, and the two classes of homogeneous bodies on the other.

Imagine four heaps, the first of needles only, the second of pieces of thread only, the third of needles and threads, and the fourth of threaded needles—the threads being knotted so as to form loops.   The first two heaps symbolise elements, say oxygen and hydrogen. The third heap symbolises a mixture, say electrolytic gas.   From this heap unlike heaps can be made by mere sorting.   The fourth heap resembles a compound, say, water, for while mere sorting alone will not resolve this heap into two dissimilar ones, yet an isolation of its constituents will be practicable after the destruction by some extrinsic agency of the continuity of the thread loops in the one case, or of the eyes of the needles in the other.

[1] See pages 144, 155.

[2] In very interesting communications to the *Chemical Society's Journal*, vol. lxi. p. 114, Picton and Linder have shown that there is a continuous series of grades of solution passing without break from suspension to crystallisable solution.

But to return to the elements which have also constituted a field for much of the more recent chemical
research. The list of elements has undergone many
vicissitudes since its first establishment. Until the
year 1811, when Davy introduced the powerful analytic method of electrolysis into chemistry, the alkalis
potash and soda, and the alkaline earths, baryta,
strontia, and lime, were regarded as elements. On
the other hand, chlorine up till this date was regarded
as a compound, being the oxide of a hypothetical
element. The establishment of the compound nature
of the alkalis and alkaline earths, and the simple nature
of chlorine was attended by important modifications
in the reigning theories of chemistry. Nitrogen was
for a time deprived of elementary dignity, being
regarded as the oxide of an unknown element nitri-
cum, and even the elementary nature of such undoubtedly simple substances as hydrogen, phosphorus,
and sulphur, was for a time called into question by
Berzelius.

It is perfectly impossible to state definitely the number of the elements; indeed, the practical proof of the
simplicity of a given matter is one of the hardest tasks
that falls to the lot of the chemist. Lavoisier's list
only contained twenty-six members, including what
are now known to be forms of energy, heat and
light, but most chemists are now agreed that the
number of unequivocal elements lies somewhere between sixty-six and seventy. This uncertainty arises
from the fact that in recent years it has been found

necessary to open a suspense list for the temporary accommodation while *sub judice* of the many new claimants for elementary distinction, and opinion is divided as to the exact number of elements which ought to be suspended. Even when two naturalists agree as to the absolute number of unequivocal elements, it does not follow that they favour exactly the same elements throughout. One may favour the advancement of an element A from the suspense to the established list, while the other may think the evidence incomplete in the case of A, but satisfactory in the case of another claimant B.

Many of the elements constituting this suspense or probationary list are not known in the free state, but only in combination with oxygen in the form of very refractory (*i.e.*, difficultly reducible) oxides. Refractory oxides are generically known as earths, and as the particular oxides under consideration are, though widely diffused, only met with in nature in very small quantities, they are usually referred to collectively as the rare earths. Many of these oxides, and therefore presumably the metals which they contain, are so similar in their chemical properties that they can neither be separated nor distinguished from one another by ordinary chemical operations or tests. It is not improbable that some of them will turn out on nicer investigation to be either variable mixtures or allotropic forms, while others *may* prove themselves compounds of already known elements and so share the fate of Bergman's "siderum," which turned out to be nothing else than iron phosphide, and Richter's "nickolanum," which was merely an impure nickel.

The most important sources of these rare and doubtful elements are the minerals gadolinite, cerite, and samarskite.

From gadolinite, Bunsen and Bahr isolated the two new earths yttria and erbia. These they regarded as the pure oxides of new elements which they called yttrium and erbium, but they did not isolate these elements. Their yttria was pale yellow in colour.

Soon after this isolation, Smith and Delafontaine separated a much darker yellow yttria from samarskite. They therefore concluded that it was impure, and succeeded in separating from it a new orange-coloured earth, terbia. This left a residue of perfectly white yttria. Hence they concluded that Bunsen and Bahr's yttria was non-homogeneous. Nor did Bunsen and Bahr's erbia prove itself a simple substance, for Marignac succeeded in resolving it into what he called true erbia and ytterbia. Nilson then found that ytterbia was non-homogeneous, and split it up into true ytterbia and the earth scandia predicted by Mendeléeff. Clève, in making a spectroscopic study of Marignac's erbia, came to the conclusion that it was not homogeneous, but was a mixture of true erbia and two new oxides, holmia and thulia. Finally, de Boisbaudran brought forward reasons for suspecting the integrity of holmia, which he ultimately resolved into true holmia and dysprosia. This little sketch does not include all nor nearly all the rare earths;

it is intended to be illustrative rather than comprehensive and encyclopædic.

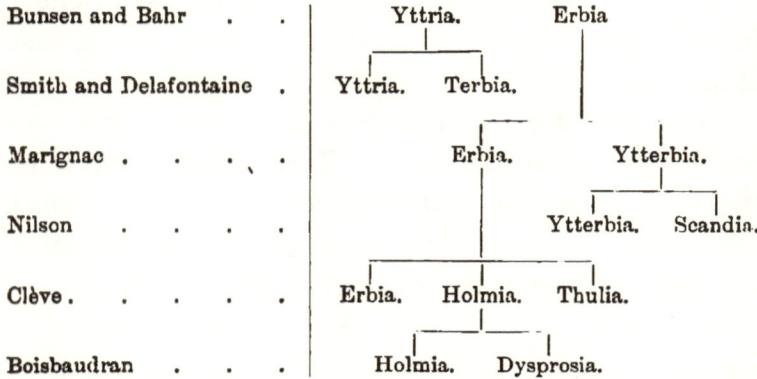

Bunsen and Bahr	Yttria. Erbia
Smith and Delafontaine	Yttria. Terbia.
Marignac	Erbia. Ytterbia.
Nilson	Ytterbia. Scandia.
Clève	Erbia. Holmia. Thulia.
Boisbaudran	Holmia. Dysprosia.

Thus Bunsen and Bahr's original alleged simple earths yttria and erbia were far from being homogeneous, as they believed; they were, according to present views, mixtures of at least seven discrete oxides.

The methods which have been chiefly used in resolving these earths are fractional precipitation of solutions of the oxides in acids by ammonia; fractional decomposition by heat of the mixed nitrates; and fractional crystallisation as already exemplified in the case of potassium chlorate and chloride.

If to a solution which contains two earths of different basicity (and all the earths differ to a greater or less degree in this property) insufficient ammonia be added to precipitate them both, then the less basic earth yields more completely to the precipitant, and can be filtered off, while the more basic constituent resists its action and remains in solution. Thus after several thousands of such fractional precipitations conducted in an orderly manner, the two earths may be completely separated from one another so far as ammonia can effect this separation. The separated products are then ready for, and should be subjected to, some other kind of fractional resolution. If they yield to no other fractional methods, then they are *pro tem.* elements.

As regards separation by fractional decomposition of nitrates, it should be mentioned that here again the nitrate formed from the less basic earth yields more easily to the decomposing action of heat than the nitrate formed from the more basic earth. Let A and B be two earths, of which A is more basic than B. Form the nitrates of these earths by acting on them with nitric acid, and then partially decompose by heat the mixed nitrates into their corresponding oxides. Take up the residue with water and filter. The residue will consist chiefly of the oxide of B and the filtrate of the undecomposed nitrate of A.

In addition to the rare earths already mentioned, the following, among others of *very* doubtful claims, are candidates for the class of simple substances:—yttrium, ytterbium, zirconium, thorium, decipium, lanthanum, samarium, cerium, praseodymium, neodymium.

The oxides of all the substances so far mentioned have been isolated, and the elements of which they are oxides have had provisional numbers assigned for their atomic weights. These numbers are necessarily affected with more or less uncertainty, for in many cases they depend on arbitrarily assumed formulæ for the oxides.

But even the admission of all the foregoing substances into the " elementary hierarchy " does not satisfy
some chemists, who, either from the spectroscopic examination of minerals from different sources, or from
the results of long-continued and varied fractionations
in the laboratory, have come to the conclusion that
most of the above alleged elements are still mixtures. These chemists have not, however, succeeded
in isolating the constituents of these mixtures;
they only find, according to their interpretation of
spectroscopic phenomena, *indications* of non-homogeneity.

Thus Crookes believes yttrium to be a mixture of
at least five elements, and Krüss and Nilson would
find at least nine elements in the once reputed element
didymium, not two, praseodymium and neodymium, as
Welsbach thinks. But it should be here emphasised
that even could these elements be isolated they would
not be distinguishable from one another chemically.
Their discrimination would demand the subtile search-
ing power of the spectroscope, and our most refined
experimental methods would fail to establish *conclusively*
any difference in their atomic weights. As yet, these
" meta-elements," as they have been called by Crookes,
are mere possibilities, not certainties, and it may be
said that their substantiation, while of great speculative and theoretic interest, would not at all influence
ordinary practical chemistry, which would still continue
to regard yttrium, for instance, not its five meta-
elements, as its working unit. It should also be

remembered in this connection that as yet spectroscopic science "is still, for want of laws, at the epoch of accumulation of facts, not of their possession." It is as if chemists were making their preconceived notions of the nature and number of the elements to fit in with spectroscopic phenomena, instead of first thoroughly establishing principles of spectroscopy, and then using these to interpret the results of their investigations on the elements.

Generally speaking, the upholders of the Periodic Law [1]—perhaps the greatest chemical generalisation of modern times—are averse to the idea of any great increase in our commonly accepted number of elements. In the tabular scheme which Mendeléeff gives as the expression of this law, only six of the rare metals are admitted, yttrium, ytterbium, lanthanum, cerium, thorium, and zirconium, and there are only, so to speak, vacancies left for a very limited number of other elements, unless indeed the new elements have atomic weights greater than that of uranium, which has the highest atomic weight (240) of all known elements. None of the above claimants would fall into this category, and it seems very improbable that many, if any, elements with atomic weights greater than 240 remain to be discovered.

This recent work on the rare earths, combined with the results of the application of the spectroscope to the investigation of the stars and nebulæ, has resulted

---

[1] See article, "Periodic Law," in Muir and Morley's new edition of *Watts' Dictionary of Chemistry*.

in modern views regarding the nature of the elements which approximate to those held by the Aristotelian school, in so far as they postulate one fundamental matter. No longer are the elements universally believed to be ultimate and independent in their nature, as Boyle and Lavoisier insisted, but the belief is growing that they consist of one fundamental matter in various stages of condensation, and that the stage of condensation is dependent on the temperature, being less the higher the temperature—a belief which reinvests the alchemistic attempts at transmutation with a certain amount of warranty. In 1815 Prout suggested that this fundamental matter was hydrogen, but the suggestion was shown to be incorrect by the classical work of Stas and others. At present the fundamental matter called "protyle," without any further attempt being made to characterise it, is, so to speak, placed lower down in the scale than hydrogen, which is itself regarded as protyle in a certain advanced stage of condensation.

Crookes regards all elements as having been gradually evolved from this protyle; the heavier elements, such as bismuth, thorium, and uranium, being the younger species; the lighter elements, such as lithium boron, &c., being the older. A consequence of the adaptation of this theory to the teaching of the periodic law, is that what we are accustomed to call elements ought really to be called "elementary groups"; that what we are accustomed to regard as the atomic weight of an element does not accurately represent the weight of each atom constituting that element, but the mean of the weights of the atoms making up a definite mass of the element, these atoms being of slightly different weights. Thus the atoms constituting a mass of yttrium have not all identically the same weight; while some, it may be the greater quantity, weigh 89 times as much as the average weight of the hydrogen atom, a few weigh 89·01, a few others again 88·99 times as much, and so on. As before stated, Crookes believes that there exist in the "element" yttrium, at least five different kinds of atoms, each constituting what he calls a meta-element. These meta-elements are regarded as protyle in five *very slightly* different stages of condensation. He is not quite decided as to whether one of these meta-elements largely preponderates over the other four as above hinted, but he rather inclines to this view. If such is the nature of the yttrium commonly accepted as an ·element, what, he asks, is there improbable in the idea that minute investigation would result in the assigning of a similar structure to the more common and familiar elements—to calcium and barium, for instance ?

If we accept this bold application of the principle of evolution to the elements, we are scarcely justified in any longer speaking of matter as inanimate. According to Crookes, the atoms of the elements are born, undergo secular vicissitudes, and die; and the rare elements are, in his opinion, strictly comparable to rare plants and animals, in that they are simply examples of special kinds of atoms that have failed to harmonise with their environment.[1]

---

[1] For further details on this abstruse and highly speculative subject, reference must be made to Crookes' original papers in the *Journal of the London Chemical Society*, vols. liii. and lv. In this connection Clerk Maxwell's article " Atom " in the *Encyclopædia Britannica* should also be read. In this article reasons are advanced for the diametrically opposed belief that the atoms of any one element, hydrogen, for instance, are eternal and identically similar among themselves, bearing the impression, so to speak, of articles stamped by machinery.

CHAPTER IV.

THE ATOMIC THEORY.

FROM the beginnings of philosophy the question of the constitution of matter [1] has exercised the minds of philosophers. Defining matter as that which occupies space, it has been asked, does matter completely fill the space its outline occupies or not? Does the gas contained in a flask completely fill the space it occupies, *i.e.*, the space defined by the sides of the flask, in the same way as to our gross senses jelly fills a cup? Or does it, on the contrary, fill the space in much the same way as apples fill a barrel, so that with sufficiently magnified vision we could plunge an exceedingly delicate style into the flask in such way that the style's point, occupying at a given instant an interstice between the individual particles of the gas, would not at that instant actually touch the gas? [2]

[1] The implications of the terms *constitution of matter* and *nature of matter* must be carefully distinguished. The ultimate nature of matter is a metaphysical rather than a chemico-physical question. For a *résumé* of the varied views held on this subject see Tait, *Properties of Matter.*

[2] It should be remarked that the question of the structure of matter is here being developed from the conventional chemical standpoint, not from the physical. The conception which most chemists, I think, have of a piece of chalk, for instance, is a collocation of molecules, each made up of hard atoms, each having the incompletely defined composi-

Stating the question otherwise; is apparently continuous matter really continuous and capable of infinitely small division without losing its individuality, or is it really discontinuous and capable of division down only to the limit marking the beginnings of discontinuity?   Are geometrical conceptions or arithmetical to underlie our theories of the constitution of matter?

The early Greek philosopher Leukippos (428 B.C.) was the first to make answer in a vague way to these questions.[1]   He decided that matter is not continuous,

---

tion x CaCO$_3$, and each separated from its neighbours by "*void.*" But physicists, whilst admitting the ultimate heterogeneousness or grainedness of apparently homogeneous matter, do not deny its perfect continuity.   The heterogeneity *may* take the form of alternations of matter and void, but it may involve only such differences as exist between solid and fluid, or between substances differing enormously in density; or such heterogeneousness as differences in velocity and direction of motion, in different positions of a vortex ring in a homogeneous liquid. Three points should be noted in this connection—(1) in discussing the constitution of matter physicists employ the terms atom and molecule in a very loose way; (2) their connotation of the term heterogeneity is quite different from that which the same term bears in Chap. III.; and (3) although in the restricted domain of chemistry we make provisional use of the conception and term atom, yet from a broadly philosophic standpoint we do not assert that the chemical atom represents the absolute limit to the divisibility of matter.   The sociologist adopts as his unit the individual, quite regardless of the anatomist's powers. The atom is the chemical individual.   See Lord Kelvin, *Popular Lectures and Addresses*, vol. i., and Maxwell's article "Atom" in the *Encyclopædia Britannica*.

[1] Later antiquity chiefly studied the atomic theory in the more elaborated form it attained under Demokritos, the pupil of Leukippos, and thus the disciple has pretty well usurped the place of the master in history.   For a very concise account of the passage of Eleaticism into Pluralism, thence into Atomism, with the fundamental changes of idea this passage involved, see Burnet's *Early Greek Philosophy*.

but is made up of numerous small separate and indivisible particles called atoms. His views, embraced and developed by the Epicurean school, have been preserved for us by the Roman poet Lucretius (B.C. 99) in his "De rerum naturâ." Here is the main argument, as presented in the poem. First comes the perfectly gratuitous assumption that reproduction—the agglomeration of scattered, invisible particles into visible bodies—is slower than decay, the breaking up of bodies into invisible particles. From this premise it follows that there must be a limit to the breakage; else the decay of infinite past ages would have left nothing visible and tangible. This limit to breakage or decay involves the idea of a least in things, which least is called an atom. Lucretius also taught that there were atoms of many different forms and weights—an infinite number of each kind; and that all change consisted in the intimate contact or separation from such contact of these different kinds of atoms. This view of matter as presented by Lucretius is more of a vague poetic cosmogony than a scientific theory. It had no numerical conciseness about it, and was altogether too indefinite to be suggestive. It bears to the atomic theory, as accepted at present, about the same relation as do the scientific dreams and guesses of a Goethe to the experimentally substantiated inspirations of a Newton.

Passing over the use which Newton made of an atomic conception of the atmosphere to make his theoretically derived velocity of sound tally with the experimental value—a physical rather than a chemical point—we do not find any serious attempt at a strictly scientific theory of the constitution of matter till we come to the year 1804. In this memorable year John Dalton first propounded his atomic theory—a theory which is so fundamental in the chemistry of to-day, that were it in any way disproved, the whole superstructure of the science would fall in hopeless and chaotic ruin.

Lord Kelvin has said that only when we can *measure* the thing we are speaking about and express it in numbers do we begin to know really anything about it. Judged by this sentiment, Dalton brought us to the beginnings of a true knowledge of atoms, for as we shall see his theory was nothing else than that of the early Greek philosophers mentioned, founded not on fancy, but on the careful observation of the facts of nature, and reduced to numerical definiteness. Let us see how Dalton was led to embrace the atomistic or limited divisibility view of matter.

The fact that water does not dissolve all gases in the same quantities seems first to have suggested to Dalton the view that gases consist of large numbers of discrete particles, but what in his opinion chiefly demanded a resuscitation of the atomic theory was undoubtedly his discovery of the law of multiple proportions taken in conjunction with the previously substantiated law of definite proportions. Hence we must first glance at these two fundamental chemical laws.

The discovery of the law of definite proportions cannot be ascribed to any particular individual; it was arrived at simultaneously by several chemists soon after Lavoisier introduced quantitative methods into the science. If A, B, &c., represent the masses in which definite homogeneous substances interact to produce new homogeneous substances represented quantitatively by C, D, &c., then this law in its most general form states that no matter what the conditions of reaction or the absolute values of A, B, C, D, &c., may be, the ratios $\frac{A}{B}, \frac{A}{C}, \frac{A}{D}, \frac{B}{C}, \frac{B}{D}, \frac{C}{D}$, &c., are all constant. A particular case is when A and B, &c., represent elements uniting to form the single compound C, and the law is generally enunciated in terms of this special case as follows: no matter the conditions of formation, or the proportions in which the constituents are taken, a given chemical compound has always exactly the same percentage composition.

Most laws (e.g., the gaseous laws of Boyle and Charles) are only true under specified conditions or between certain narrow limits, but the law now under observation is as far as we can tell an absolute and unconditional one. Working with the haloid salts of silver —preparing them in many different ways and from different proportions of the ingredients—Stas proved the law accurate to 1 part in 10,000 parts,[1] an accuracy

[1] An ordinary chemical balance is capable of an accuracy of 1 in 10^6, and Miller in his elaborate construction of the standard pound attained an accuracy of 1 in 3×10^6, but in Stas' experiments errors considerable in magnitude when compared with those involved in mere weighing, are incidental to the operations of filtering, washing, incinerating, &c. An accuracy of 1 in 10,000 is about that attainable in a simple measurement of length.

of high order when we recall the diverse operations involved.

Yet the law of definite proportions has had its vicissitudes. In the beginning of this century Berthollet denied its truth as above enunciated, asserting that the composition of a body might vary within certain limits, this variation being dependent on the masses in which its constituents were allowed to react [1]—just as recipes with slight quantitative differences will give to all intents and purposes the same culinary product. In this mistaken view he was successfully opposed by Proust, who was led to a warm espousal of the law by the following experiment. Proust analysed native malachite and found that it gave 71·9 per cent. of cupric oxide. He then dissolved the mineral in acid and reprecipitated the malachite in an amorphous form. This artificial product gave on analysis 71·9 per cent. of cupric oxide. Hence Proust concluded that no matter whether malachite is formed in the crystalline condition in the earth's crust, where it is subjected to the influence of mighty secular processes of which we know next to nothing, or whether it is formed artificially and suddenly in the laboratory, it has in each case identically the same composition. He then showed that the alleged variable nitrate of mercury which Berthollet brought forward to vindicate his attitude, was not a

---

[1] The influence of mass in causing variations in the relative quantities of the products of a change (not of their individual compositions as Berthollet supposed) lies at the foundation of the study of chemical affinity in the modern acceptation of this term. See Pattison Muir, *Principles of Chemistry.*

homogeneous body but a mixture of mercurous and mercuric nitrates, in proportions varying according to the relative masses of mercury and nitric acid brought together. Thus arose the distinction between homogeneous substances and non-homogeneous substances or mixtures detailed in Chap. III.

Passing on now to the law of multiple proportions. Before Dalton's time the fact had been recognised that a given element A will combine with another element B in two or more proportions to form two or more compounds, for each of which the law of definite proportions is true. But up to the year 1803, owing to the customary method of stating the results of analysis, no regularity had been detected in the values of the combination masses of A and B as between compound and compound. Dalton analysed the two compounds of carbon and hydrogen, methane and ethylene; as also the two compounds of carbon and oxygen, carbonic oxide and carbonic anhydride, with these results—

ETHYLENE.[1]

C = 85·71 p.c.   H = 14·28 p.c.

or approximately

$$\frac{C}{H} = \frac{6}{1}$$

METHANE.

C = 75 p.c.   H = 25 p.c.

or approximately

$$\frac{C}{H} = \frac{6}{2}$$

CARBONIC OXIDE.

C = 42·86 p.c.   O = 57·24 p.c.

or approximately

$$\frac{C}{O} = \frac{6}{8}$$

CARBONIC ANHYDRIDE.

C = 27·27 p.c.   O = 72·72 p.c.

or approximately

$$\frac{C}{O} = \frac{6}{16}$$

---

[1] The percentages here given are not the values actually obtained by Dalton.

No regularity appears between the valves for the two hydrides on the one hand, or the two oxides on the other, until we consider a constant mass of either of the common elements in the two cases, instead of expressing the results in the conventional percentage form.   It is then noticed that the amount of hydrogen combined with six unit masses of carbon in marsh gas is *exactly* [1] double (not $2\frac{1}{3}$ times, nor $1 \frac{99}{100}$ times, but exactly double) the amount of hydrogen combined with six unit masses of carbon in ethylene.   Similar observations apply to the oxides of carbon, and indeed to all cases where two elements combine in different proportions to form different compounds.   In every case the following regularity is observed.   If an element A combines with an element B in two or more proportions to form two or more compounds, and if in these compounds we consider a fixed mass of either of the elements, say A, then the masses of B combined in the several compounds with this fixed mass of A, are in general so related that the compounds richer in B contain of B masses expressible as whole multiples of the mass of B, contained in the compound poorest in B.   This is Dalton's law of multiple proportions. [2]

[1] This exactness is never of course attained in practice, on account of unavoidable errors of observation, but it is more and more nearly realised as we perfect our analytic methods.   Hence the assumption is warranted that the relations in question are in reality absolutely exact. The mathematical conception of a limit is here implicated.

[2] Some apology for this diffuse enunciation seems called for.   In the majority of chemical text-books, the word multiple is never once introduced in the enunciation of the law of multiple proportions, its place

What explanation can be adduced of these laws?[1]
Dalton answered, the atomic theory of matter. Admit,
he said, that simple matter of every kind consists of
little indivisible particles called atoms, that the atoms
of one and the same kind of matter have always exactly
the same weight and the same properties, but that the
weight and other properties of the atoms differ from
substance to substance; that chemical combination
consists in the coming together into intimate contact
of definite numbers of simple atoms, to form definite
compound atoms which in turn are all exactly similar
for one and the same compound, bearing, so to speak,
the impress of goods stamped by machinery; then,
admitting all this, the fundamental laws of chemistry

being taken by the vague, elastic, yet polarised term "simple ratio."
In the statement of the law given in the text the attempt has been
made to remedy these defects, even at the expense of brevity.

[1] In addition to the two laws given in the text, many add a third—
the law of reciprocal proportions, or the law of combining weights.
As a matter of fact, the perception of the numerical relations embraced
by this law did not antedate and pave the way for the atomic theory,
but this theory established, certain numerical results necessarily following from it were grouped together under one or other of the above
titles, thus:—if AB and AC represent quantitatively two compounds,
then any compound of B and C will be represented quantitatively by
$nBmC$, where n and m are whole numbers, and A, B, C, the combining weights of the elements. Interpreted in terms of the atomic
theory, this law may be regarded as stating that the atoms of a given
element have a fixed and unalterable mass in all the compounds that
element forms. AgI and $AgIO_3$ might each have fixed compositions without the ratio $\frac{Ag}{I}$ being identical in the two cases as theory
demands. Stas, in proving to an accuracy of 1 in 10^7 that the
ratio $\frac{Ag}{I}$ is the same in both compounds, has placed the atomic theory
on a very exceptional experimental basis.

follow as necessary consequences.[1] It is quite easy to see how the law of definite proportions follows; let us examine in detail the sequence of the multiple proportion law.

Suppose matter to be non-atomic, to be infinitely divisible and absolutely continuous. Let the symbol 6 represent a definite mass (expressed numerically by 6) of carbon on this supposition; and the symbol ▨ represent under the same conditions a mass of hydrogen, numerically expressed by 1. Then ethylene might be represented as [6▨]. Now, analysis informs us that methane contains relatively more hydrogen than does ethylene. According to the above supposition, the mass of hydrogen combined with 6 unit masses of carbon in methane might quite easily be $1\frac{1}{7}$, $1\frac{1}{9}$, $1\frac{15}{16}$, or $2\frac{70}{71}$, &c. times the mass of hydrogen combined with 6 unit masses of carbon in ethylene; for hydrogen is jelly-like and structureless, and we can imagine a

---

[1] Stallo divides the phenomena of chemical change into three classes. The first "embraces the persistence of weight and the combination in definite proportions; the second, the changes of volume and the evolution or involution of energy; and the third the emergence of a wholly new complement of chemical properties." He asserts that the atomic hypothesis is in no sense an explanation of phenomena of the second and third classes, nor does it fully explain those of class i. in the sense of generalising them and reducing many facts to one. It accounts for them by simply iterating the observed fact in the form of an hypothesis. He admits, however, "the merits of the atomic hypothesis as a graphic or expository device—as an aid to the representative faculty in realising the phases of chemical or physical transformation." See *Concepts of Modern Physics*, chap. vii.

quantity of it of any mass whatever. Given ⬚6▨

as representing ethylene, the composition ⬚6▨ would,

a priori, be quite as probable for methane as would

⬚6▨ .

But now let us take the atomic theory, and for the sake of present argument assume that ethylene consists of 1 atom of carbon weighing 6 units, and 1 atom of hydrogen weighing 1 unit. Then, since methane contains more hydrogen relatively than ethylene, it follows that per 1 atom of carbon methane must contain 2, 3, 4 . . . . . n atoms of hydrogen; for atoms are the indivisible units of chemistry. Since all atoms of the same simple substance have exactly the same weight, the law of multiple proportions necessarily follows; the amount of hydrogen per given amount of carbon must necessarily be in methane a whole multiple of its value in ethylene.

To sum up the whole matter; the existence of a multiple law according to the non-atomic theory of matter is a possibility among an infinity of other equally plausible possibilities; the odds against its existence are infinitely great. But of an atomic theory of matter, a multiple law is an absolutely necessary consequence. That Dalton clearly saw this constitutes, perhaps, his greatest claim on the memory of posterity.

We must now follow Dalton in his attempt to find the weights[1] of the atoms of elementary bodies. He did not, of course, hope to determine the absolute weights of the atoms — a determination of which even the advanced chemistry of to-day is incapable. All he attempted to do was to find the relative weights of the atoms. He made the hydrogen atom his standard, and arbitrarily called its weight one; all other atomic weights were stated in terms of this standard. Thus the statement—the atomic weight of oxygen is 16— simply means that the atom of oxygen is sixteen times heavier than the atom of hydrogen, whose absolute weight expressed, say in a fraction of a milligram, is unknown.

As an aid to the foundation of a system of atomic weights, Dalton framed a series of empirical rules. We cannot dissect out the train of reasoning which culminated in these rules. They are merely the expression of Dalton's own preconceived notions—mere guesses, limited only by the condition that compound atoms are of simple rather than of complex structure, being as a rule made up of very small numbers of elementary atoms. Rule 1. If only one compound of two elements can be obtained, the compound must be assumed a binary[2] one, unless there be good cause for

---

[1] In some recent text-books, *atomic mass* replaces the term *atomic weight*. In view of the strict proportionality which exists between mass and weight, and in view of the hold the terms weighing and weight have in the speech of everyday life, this innovation seems to me unnecessary.

[2] A binary compound, according to Dalton, is one whose "com-

some other conclusion. Rule 2. If two compounds of two elements exist, then one is a binary, and the other a ternary, compound. Rule 3. If three compounds of two elements exist, then one is a binary, and the other two are ternary, compounds. Rule 4. If four compounds of two elements are known we should expect two of them to be ternary, one to be binary, and the fourth to be quaternary. Other rules follow, dealing with the specific gravities of binary, ternary, &c. compounds; but for these, and indeed for Dalton's work on the atomic theory as a whole, the reader is referred to Ostwald's *Klassiker der exakten Wissenschaften*, No. 3, or to the *Alembic Club Reprints*.[1]

To illustrate the application of these rules, we may take the case of the atomic weight of oxygen. To Dalton only, one compound of hydrogen and oxygen was known, viz., water. Hence by Rule 1 the compound atom of water consisted of one atom of hydrogen in combination with one atom of oxygen. Now it is found by actual experiments that about 8 grams of oxygen combine with 1 gram of hydrogen to form about 9 grams of water; or, what is the same thing, that $\frac{8}{10^{23}}$ milligrams of oxygen combine with $\frac{1}{10^{23}}$ mg. of hydrogen to form $\frac{9}{10^{23}}$ mg. of water. Let us sup-

pound atom" contains only two simple atoms. In the "compound atom" of a ternary body there are three atoms, and so on.

[1] These brochures are more generally attainable than the *Memoirs of the Literary and Philosophical Society of Manchester*, in which Dalton's papers originally appeared.

F

pose, simply for the sake of clearness, that a hydrogen atom weighs $\frac{1}{10^{23}}$ mg.[1]

$\frac{1}{10^{23}}$ mg.   H combines with $\frac{8}{10^{23}}$ mg. O.

or 1 atom H   ,,   ,,   ,,

but, by Rule i., 1 atom H   ,,   ,,   1 atom O.

Therefore 1 atom O weighs $\frac{8}{10^{23}}$ mg.

So if we agree to call the atomic weight of hydrogen one, the atomic weight of oxygen will equal eight. In a perfectly similar manner Dalton arrived at the result; atomic weight of nitrogen $= 4\frac{2}{3}$. But 16 ($= 8 \times 2$) and 14 ($= 4\frac{2}{3} \times 3$) are at present regarded as the approximately correct values for the atomic weights of oxygen and nitrogen respectively. Why these numbers are adopted in preference to Dalton's will not be apparent till we have treated of Avogadro's hypothesis.

When we attempt to fix by Dalton's rules the atomic weight of such an element as carbon which combines with hydrogen, oxygen, &c., in more than one proportion, we find ourselves on the horns of a dilemma. According to Rule 2, either methane or ethylene is a binary body, but the rule does not indicate any means of assuring ourselves which is the binary body. If ethylene be assumed binary, methane will be ternary. Ethylene will be CH, methane CH_2, and the atomic weight of carbon will be 6. If, however, methane be

---

[1] This very rough approximation is derived from data furnished by the kinetic theory of gases.

assumed binary, ethylene will be ternary. Under this assumption, the formula for methane will be CH, that for ethylene C_2H, and the atomic weight of carbon will be 3.

Which of these two values for carbon are we to select? There is nothing at all in Dalton's rules to guide us in such cases as these. Failing inspired arch-chemists, one is as justified from the premisses in maintaining that the atomic weight of carbon = 3 as another is in championing the value 6.

This ambiguity in Dalton's rules led to great differences of opinion respecting the atomic weights of several of the elements, and ultimately the confusion became so great that many advocated the renunciation of the atomic theory altogether. The light which it indisputably shed on some points did not in the opinion of many compensate for the fundamental uncertainties with which it was hampered, and which it was powerless to resolve. It is, I think, worthy of note that the very cases which in the first instance suggested to Dalton the law of multiple proportions and the atomic theory were exactly those cases in which his elaborated system was found wanting, and which led to its unpopularity and temporary rejection.

This indefiniteness, which trammelled the young atomic theory, was finally resolved by the knowledge accruing from a careful study of the physico-chemical properties of gases, to which we now turn.

In 1805, Gay-Lussac and Humboldt, investigating the constancy of the amount of oxygen in the atmosphere, employed an analytic method first proposed by Volta. This method, which has become classical, consists in mixing the air with hydrogen in a eudiometer, exploding the mixture by an electric spark, and then from the contraction which ensues calculating the amount of oxygen present. To apply this method, an accurate knowledge of the contraction which occurs when liquid water is formed from hydrogen and oxygen is obviously essential. After very careful experiments, Gay-Lussac and Humboldt found that one volume of oxygen combines with two volumes of hydrogen to form a drop of water, whose volume in comparison with the volumes of the gases exploded is in general negligible.

2 vols. H + 1 vol. O = water with practically negligible vol.

Hence it follows that 3 volumes of the properly mixed gases contract to zero volume on explosion. In other words, suppose 2 cubic feet of hydrogen and 1 cubic foot of oxygen (measured at atmospheric pressure), placed in an air-tight vessel[1] and then exploded, the pressure in the vessel would sink from 15 lbs. per square inch to nearly zero value. I say *nearly*, for the water formed would vaporise and exert a small pressure (or as it is currently but unfittingly called, a tension) dependent as regards magnitude on the tempe-

---

[1] This experiment was actually performed by Cavendish, who was the first to determine the composition of water by volume (see Chap. II. p. 53). Cavendish introduced the gases, hydrogen and oxygen, mixed in the proper proportions into a vacuous globe, exploded the mixture, and found that the vacuum was re-established, so that several successive charges and explosions could be effected with one initial-exhaustion of the globe.

rature. From all of this we conclude that the amount of oxygen in a mixed gas is equal in volume to one-third the contraction caused by sparking after an excess of hydrogen has been added.

Supposing the water formed from 2 vols. hydrogen and 1 vol. oxygen is changed into the gaseous state by heating it above 100° C. so as to form super-heated steam; or supposing the whole experiment is conducted throughout at a high temperature, so that the water formed never condenses, what volume-relation would the water gas or steam bear to the volumes of hydrogen and oxygen which formed it? In answer to this inquiry, Gay-Lussac and Humboldt found that the volume of steam formed is exactly equal to the volume of the hydrogen (or what is the same thing, to twice the volume of the oxygen) exploded.

Here Gay-Lussac and Humboldt met with a quantitative fact which astonished them by reason of the extreme simplicity of the quantities involved. Two cubic feet of hydrogen combine with *exactly* one cubic foot of oxygen, not with ·9 or $1\frac{1}{5}$, but according to Gay-Lussac and Humboldt with *exactly* 1 cubic foot, of oxygen to form, not 2 $\frac{19}{1000}$, but *exactly* 2 cubic feet of steam at the same temperature.[1]

[1] Although we assert in the text the exactness of the ratio 2 : 1, yet it seems well nigh impossible to prove this exactitude experimentally. Scott and E. W. Morley, employing all the refinements of modern methods, have repeated Gay-Lussac and Humboldt's experiment with the following result :—

2·00245 vols. of hydrogen combine with 1 vol. of oxygen (Scott).
2·0023 ,, ,, ,, ,, (Morley).

However, in all the theoretical deductions which follow, any slight

The question suggests itself, does a similar simple relationship in the reacting volumes manifest itself in other cases of gaseous combination ?   It does.

1 vol. of hydrogen combines with exactly 1 vol. of chlorine to give 2 vols. of hydrochloric acid gas.

1 vol. of nitrogen combines with exactly 3 vols. of hydrogen to give 2 vols. of ammonia gas.

2 vols. of nitrogen combines with exactly 1 vol. of oxygen to give 2 vols. of laughing gas.

1 vol. of carbonic oxide combines with exactly 1 vol. of chlorine to give 1 vol. of phosgene gas.

Several other instances might be adduced of Gay-Lussac's Law of Volumes, which states that in homogeneous gaseous reactions (*i.e.*, reactions in which all the factors and products of the change are gaseous) all the volumes involved are in such simple relationship that the ratios can in every case be expressed in terms of the first six digits.[1]

Just as the law of multiple proportions called forth an explanation, so it was not long before the why and wherefore of Gay-Lussac's law was under discussion.   The outcome of this discussion was the conjec-

---

deviations from simplicity which may possibly exist in the ratios of the combining volumes of gases will be disregarded, and the law of volumes will be accepted as strictly true.

[1] It should be noticed that when the factors of a homogeneous gaseous reaction are elements, the volume of the resulting compound is, with very few exceptions, always 2, provided the volume equation be throughout reduced to its simplest terms.   The formation of phosphine from its elements furnishes an example of the exceptions referred to.

$$P_4 + 6H_2 = 4PH_3$$
· 1 vol. + 6 vols. = 4 vols.

ture [1] accepted by several chemists, but first definitely
enunciated by Berzelius, that equal volumes of gases
contain equal numbers of atoms, and hence that the
atomic weights of gases are proportional to their
specific gravities.[2] In view of the identical behaviour,
qualitatively and quantitatively, of different gases when
subjected to pressure and temperature changes, this
conjecture seemed very plausible. The laws of Boyle
and Charles seem absolutely to demand some such identity in mechanism of the various gases as is implied in
Berzelius' conjecture. But apart from this consideration, it cannot be denied that the human mind has a
peculiar and inherent bias for the uncomplicated, and
it is probable that the extreme simplicity of the explanation was one of the most potent factors in securing for
it a general acceptance.

Dalton and his school, however, absolutely refused to
accept Berzelius' interpretation of the law of volumes,
maintaining that the atomic weights could be determined only from " the ponderable relation of elements

[1] Berzelius' interpretation of the Law of Volumes seems scarcely to
merit the title of an hypothesis, which " is any supposition we make in
order to endeavour to deduce from it conclusions in accordance with
facts which are known to be real."

[2] Since the specific gravities of oxygen and nitrogen, referred to
hydrogen as standard, are respectively 16 and 14, it follows that these
values were adopted by Berzelius for the relative atomic weights of
these elements. It should be remarked that at first Berzelius regarded
the data derived from the law of volumes as belonging to what he
called "volume atoms" or "elementary volumes." These he conceived
of as something fundamentally distinct from Dalton's atoms. After a
time, however, he returned to the Daltonian conception of atoms and
applied to these the results following from his hypothesis. See Wurtz,
The Atomic Theory, pp. 43–48.

in combination," *i.e.*, from purely analytic data.   As an argument against this interpretation, those cases of gaseous combination which occur without change of total volume were adduced.

1 vol. hydrogen combines with 1 vol. chlorine, forming
2 vols. hydrochloric acid gas.

Assume that the volume of hydrogen considered is so small that it contains only one atom ; then, accepting Berzelius' conjecture, the equal volume of chlorine with which it combines will contain only one atom, and the double volume of hydrochloric acid gas formed will contain two compound atoms.   Now each of these compound atoms must contain *at least* one hydrogen atom and one chlorine atom.   Therefore the two compound atoms together must contain *at least* two atoms of hydrogen and two atoms of chlorine.   But we only started with one atom of hydrogen and one atom of chlorine.   Therefore matter has been created — an absurd conclusion in view of all the proofs that exist of the conservation of matter.[1]

Berzelius was bound to admit the justness of this *reductio ad absurdum*, and in order to meet it emphasised the fact that his so-called hypothesis only extended to the elementary gases, and not to compound gases and

[1] A neat but indirect proof of the principle of the conservation of matter is furnished by the constancy of the length of the year.   This depends on the masses of the earth and sun.   If the masses of the earth were continually changing by reason of the chemical changes taking place, then some alteration in the length of the year would have been produced within historic time.

vapours. In this form he continued to make a necessarily restricted use of it for the determination of atomic weights. Of Berzelius' work in this direction we shall have occasion to speak later.

Just about this period in the history of the atomic theory (1813), an Italian chemist, Avogadro, pointed out that the whole of the differences between Dalton's and Berzelius' attitudes towards the law of volumes would disappear if chemistry would but admit into its philosophy a new order of particles of a higher grade of organisation than the atoms. These particles he called molecules, and postulated that all the molecules of the same substance are identically similar, and in general consist of an assemblage of atoms, even in the case of simple gases. Heretofore a radical distinction had existed between the constitution of a simple, and that of a compound, gas. A mass of oxygen was pictured as an assemblage of atoms, each with perfect freedom, and completely independent of its neighbours. In a compound gas, such as hydrochloric acid, on the other hand, the particles enjoying this individuality and freedom were not single atoms—else the gas had been a mere mixture of hydrogen and chlorine—but atom complexes. Each atom of hydrogen kept perpetual company with an atom of chlorine, the combination forming a discrete and independent particle of the gas.

Avogadro asserted that this distinction between the simple and compound gases was unwarranted. He regarded both hydrogen and chlorine as, in a sense,

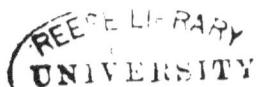

compounds; the one was to him hydride of hydrogen, the other chloride of chlorine. The freely moving and independent particle of hydrogen gas was not a single atom, it was an atom complex. The only difference between a simple and a compound gas is that the atom complex of the former is made up of similar, that of the latter of dissimilar, atoms. Having premised so far, Avogadro then concluded that equal volumes of all gases, simple and compound, contain under similar conditions of temperature and pressure the same number of molecules. This conclusion is generally known as Avogadro's "hypothesis."[1] As will be readily noticed, it is nothing else than Gay-Lussac's conjecture with one word changed—*molecules* replaces *atom*.

In 1814, Ampère independently came to precisely the same general conclusion with regard to the structure of gases, simple and compound.

It is easy to see that those instances of gaseous combination wherewith Dalton combated Berzelius' interpretation of the law of volumes involve no inconsistencies when interpreted in the light of Avogadro

[1] Avogadro's "hypothesis" is sometimes erroneously stated in the form :—All gaseous molecules under like conditions have the same size. What is really meant is that all molecules have the same sized spheres of action, which they occupy and dominate to the exclusion of other molecules. A compact square of fifty soldiers armed with rifles would, in a certain sense, dominate the same extent of country as twenty-five soldiers armed with the same rifles, although the size of the actual squares would be different in the two cases. It may here be stated that Avogadro's generalisation is *strictly* true only for perfect gases. The ratio of the number of molecules in equal volumes of oxygen and hydrogen at ordinary temperatures and pressures is about 100,020 : 100,000.

and Ampère's views. The volumetric relations of hydrochloric acid gas merely prove that the molecules of hydrogen and chlorine must contain at least two atoms each.[1]

Yet the times were not ripe for this great generalisation which to-day, under the title Avogadro's law, stands the very foundation and framework of theoretical chemistry. It attracted but little attention at the time of its birth, and soon fell into an oblivion from which

---

[1] The general introduction of these conceptions of the structure of elementary gases at a later period in the history of chemistry, threw a new light on the phenomena of substitution which for a time enjoyed a special prominence. In the formation of a halogen derivative of a hydrocarbon, it had up to date been necessary to ascribe different *rôles* to the admittedly perfectly similar atomic units of the halogen. For instance, in the formation of monochlor-methane, it was held that an atom of chlorine first replaced an atom of hydrogen in the hydrocarbon—

$$CH_4 + Cl = CH_3Cl + H,$$

and then a second atom of chlorine combined with the liberated hydrogen atom—

$$H + Cl = HCl.$$

In other words, substitution was regarded as a complex chemical change taking place in two stages, the chlorine atom in each stage playing a different chemical *rôle*. But when it was admitted that the smallest portion of chlorine entering into chemical reaction is a molecule containing two atoms, then halogen substitution was seen to be a pure case of double decomposition, impossible of resolution into two consecutive stages, and recalling in its main features exactly the substitution brought about by the action of compound bodies such as nitric acid.

The two reactions

$$CH_3H + Cl.Cl = CH_3Cl + HCl.$$
$$C_6H_5H + OH.NO_2 = C_6H_5NO_2 + HOH.$$

were seen to be quite analogous. It was no longer necessary to ascribe different *rôles* to the smallest individual particles of the halogens.

it was rescued some forty years afterwards by Gerhardt and Laurent.

In the year 1818, Dulong and Petit, experimenting on the specific heats of the elements in a solid state, discovered a most striking numerical relation between these values and the atomic weights of the respective elements. The numerical value of the atomic weight of an element multiplied by the value of its specific heat gave, in general, a constant product equal to about 6·25. The limitation "in general" is advisedly introduced, because certain of the current atomic weights did *not* satisfy this relation. On the strength of the comparatively large number of atomic weights which did satisfy their law, Dulong and Petit pronounced inaccurate all those which did not, thus virtually asserting the universality of their law, and establishing it as a powerful instrument for atomic weight determinations. They recognised the arbitrariness of the methods of determining atomic weight then in vogue; for these made the choice of one particular value out of a series, bearing a simple multiple relationship to each other, a matter of, we might almost say, individual taste.

In a small annex to a certain kitchen in the city of Stockholm, Dulong and Petit's law was warmly welcomed. In this kitchen (which has been referred to as one of the magnetic poles of the chemical world), assisted by his cook and equipped with culinary utensils rather than with what we now understand by apparatus, worked one whose name, Johann Jacob Berzelius

(1779–1848), is one of the greatest on the honour rolls of science. From the very beginning of his scientific career, Berzelius interested himself chiefly in atomic weight determinations. To convert his analytic results into values for the atomic weights, he at first made use of (1) certain rules recalling in their arbitrariness those of Dalton, (2) the law of volumes already mentioned, and (3) his so-called oxygen law.[1] Afterwards he employed in addition to these (4) the law of Dulong and Petit, and (5) the generalisation of Mitscherlich, which has of late years proved itself the very reverse of general.

At the outset, Mitscherlich believed that the correlation between crystalline form and composition was such, that a mere equality in number of the simple atoms in the compound atoms of two substances, necessitated an identity in the crystal forms, or an isomorphism of the two substances; and *vice versâ*, that isomorphism necessarily existed between substances whose compound atoms were built up by the same number of simple atoms. The discovery of polymorphism (the crystallisation of one and the same substance in different forms), however, compelled the

---

[1] Berzelius regarded salts as dual compounds of acid oxide or negative constituent, and basic oxide or positive constituent. Sulphate of soda was written $\overset{+}{Na_2O}\ \overset{-}{SO_3}$. The oxygen law stated that in all the salts of a given acid the amount of oxygen in the negative constituent bore a constant ratio to the amount present in the positive constituent. This law had been previously recognised by the German chemist Richter, who, however, had expressed himself rather obscurely on the point.

admission that atomic complexity alone was not the whole explanation of isomorphism, but that the arrangement of atoms must be taken into account. Further, the fact had to be admitted that in the majority of cases the forms of crystals of similarly constituted bodies are only approximately the same, not absolutely identical. Indeed, absolutely perfect geometric isomorphism is only found in the cases of bodies crystallising in the cubic system. Hence Mitscherlich's generalisation in its final form was not the clear-cut and unequivocal statement that it was in its original enunciation. The final statement ran as follows:—If the compound atoms of two or more *chemically analogous* bodies be composed of the same number of simple atoms (no matter the nature of the latter), then the crystals of these bodies will have identical *or nearly identical* forms.[1]

---

[1] Recent research has adduced many exceptions to this elastic generalisation, both in its direct and converse forms. The three dinitrobenzenes belong to the same chemical type, and have all the same atomic complexity; yet they have so little analogy of form that it would be an obvious overstretching of the term to call them iso-morphous. The exceptions to the generalisation in its converse form are grouped together under the titles isogonous or homeomorphous bodies. Such bodies, while differing greatly in chemical behaviour and even in molecular complexity, are nevertheless isomorphous, *e.g.*, $PbCl_2$ and $Sn(CH_3)_2Cl_2$ are isomorphous, as are also $KHSO_4$ and $KAlSi_3O_8$. It is clear throughout that the term isomorphous, which was originally a definite and definable term, is now indefinite and undefinable. What degree of similarity in the geometrical forms of crystals is necessary in order that the crystals fall in the category, isomorphous, is a question which chemists are undecided about. They prefer to judge of isomorphism by a series of chemico-physical tests rather than by purely crystallographic considerations. Two substances, A and B, crystallise in the same system with nearly the same forms

So marvellously did Berzelius balance probabilities, so carefully did he make use of analogy, so skilfully did he manipulate his roughly improvised apparatus, that his final table of atomic weights (1826), when reduced to the hydrogen standard,[1] shows remarkable agree-

and angles. Are they isomorphous? If they show the same cleavages, similar thermal conductivities, similar etched figures, approximately equal specific volumes ; if a crystal of A grows regularly in a solution of B, or *vice versâ ;* if a crystal of A causes crystallisation in a supersaturated solution of B, or *vice versâ ;* and if mixed solutions of A and B give homogeneous mixed crystals, then the answer is most decidedly yes. It is not a sufficient criterion of isomorphism that one only of these conditions is satisfied, just as an element cannot in general be classified as a metal or a non-metal from the investigation of a single property. For a good account of isomorphism, regarded as a branch of morphotropy (the general study of the inter-relation of chemical composition and crystalline form apart from considerations of similarity in the latter), see Hutchinson's article, "Isomorphism," in *Watts' Dictionary of Chemistry,* vol. iii. ; also Mendeléeff, *Principles of Chemistry,* vol. ii. p. 7.

[1] Berzelius regarded oxygen as the most important chemical element —the pole of chemistry. He therefore adopted the oxygen atom as the basis of his atomic weight system, giving it the arbitrary value 100. Modern chemistry has rather favoured Dalton's choice of the hydrogen atom as standard atom with the arbitrary value 1. Quite recently, however, the question of making oxygen the standard element again, assigning it the arbitrary atomic weight 16, has been much discussed. There is much to be said in favour of the return. The values of a large majority of the atomic weights involve an accurate knowledge of the atomic weight of oxygen. Unfortunately the determination of the atomic weight of oxygen referred to $H = 1$ is an extremely difficult chemical task, and new results differing often in the first decimal place are continually demanding recognition. Every adoption of a new value necessitates the alteration of the atomic weights of all those elements whose oxy-compounds have furnished the necessary analytical data. If $O = 16$ were universally adopted as the standard, the only atomic weight that need be affected by new data for the composition of water would be that of hydrogen ; and for all *practical* purposes the changes it would undergo might with safety be overlooked and the value 1 steadily adhered to. A small change in the accepted value of the

ments with the values for the atomic weights current to-day. To this statement we must, however, make three well-defined exceptions, viz., potassium, silver, and sodium. Lacking the data for the application of the law of specific heat in these three cases, values almost exactly twice too great were assigned to the atomic weights of these elements.

I would emphasise the fact that Berzelius made use of no one universally applicable guide in constructing his system of atomic weights. He applied to the results of analysis sometimes one, sometimes another, of the five methods above given, and in cases of doubt he selected with something akin to inspiration. One is almost tempted to say that Berzelius was lucky.

Despite the great renown of Berzelius as an analytical

atomic weight of oxygen may involve quite a large change in the values of other atomic weights, or it *may* involve none at all. The following approximate values necessary for the determination of the atomic weight of barium by Struve's method are instructive. Struve deduced the atomic weight of barium from the ratio $BaCl_2 : BaSO_4 :: 100 : 112.1$, assuming values for the atomic weights of Cl, O, and S.

	O = 16	O = 15.96
(1) Molecular weight of KCl deduced from reduction of $KClO_3$	74.52	74.33
(2) Atomic weight of Ag deduced from ratio Ag : KCl	107.8	107.4
(3) Molecular weight of AgCl from the reduction of $AgClO_3$	142.9	142.5
(4) Atomic weight of Cl (= mol. wt. AgCl − at. wt. Ag)	35.1	35.1
(5) Atomic weight of S from ratio Ag : Ag_2SO_4	31.1	30.9
(6) Atomic weight of Ba	135.6	132.66

An initial error in the atomic weight of oxygen of $\frac{1}{4}$ per cent. is multiplied up into a percentage error of 2.2 in the atomic weight of barium as determined by Struve.

chemist of the highest rank, many refused to accept his estimates of the atomic weights, not only because they differed in many instances from those upheld by the school of Dalton, but also on account of the discovery of inaccuracies and inconsistencies inherent in the system itself.

Using Berzelius' value for the atomic weight of carbon, 12·2, many anomalies had been noticed in the results of organic analysis. The atomic weight of this element was therefore redetermined by Dumas and others, with the result that the Berzelian value was found ·2 too high. This discovery not only shook the confidence of some of Berzelius' disciples in their master, but was made by his opponents the occasion of heaping all manner of sarcasm and unjust criticism on the great Swedish chemist.

But perhaps Dumas' work on the vapour densities of elements which are solid or liquid at ordinary temperatures (1827) did more to bring the Berzelian system of atomic weights into discredit than did the detection of a $1\frac{2}{3}$ per cent. error in the atomic weight of carbon. Berzelius had arrived at the values $Hg = 200$ and $P = 31$ for the atomic weights of these elements by a combination of some of the methods above indicated. But Dumas found that the vapour of mercury is only 100 times heavier than that of hydrogen under the same temperature and pressure conditions; while the vapour of phosphorus is 62 times heavier than that of hydrogen. Therefore if, as Berzelius had maintained, equal volumes of gases contain equal

G

numbers of atoms, then the atomic weights of mercury and phosphorus must be respectively 100 and 62. If, on the contrary, Berzelius' values were the true ones, then his interpretation of the law of volumes—an interpretation lying at the very foundations of his atomic weight system—must necessarily lack generality. The atomic weights of mercury and phosphorus being respectively 200 and 31, a given volume of mercury vapour can only contain half as many atoms as the same volume of hydrogen; while a given volume of phosphorus must contain twice as many atoms as the same volume of hydrogen.

The atomic theory now enters upon the most troublous period of its career. So many different values for the atomic weights of the elements were competing with each other for general acceptance, and so vanishingly small did the chances of any universal agreement on the subject appear, that it was proposed to do away with the atomic theory and its attendant uncertainties altogether, and to return to a system of constants (with suitable notation) for the elements, which constants, being simply the numerical expressions of ascertained facts, could involve no doubt and admit of no uncertainty.

This new system, which was to bring peace and prosperity in its wake, was called the equivalent system, and was pioneered by Wollaston. According to Wollaston the symbol of an element was to represent that mass thereof which combines with unit mass of hydrogen —a pure number expressing the result of an experiment, nothing more.[1] The numerical value of this mass he called the equivalent number, or simply, the equivalent of the element. Just as those quantities of acids which neutralise, *i.e.*, combine with, a fixed quantity of base are equivalent, so those quantities of the elements which combine with a fixed mass of hydrogen were also regarded as equivalent.

Yet this new system, seemingly so simple and unequivocal in its inception, soon had to encounter difficulties as great as any that had ever beset the atomic theory. Only a few of the elements could be made to combine with hydrogen directly to form hydrides; how then were the equivalents of the remaining elements to be determined? One unit mass of hydrogen combines with $35\frac{1}{2}$ unit masses of chlorine, and chlorine can be made to combine with nearly all the elements that do not form hydrides. Here then seemed a way out of the difficulty; the equivalent of an element was taken to be that mass thereof which combines with unit mass of hydrogen, or $35\frac{1}{2}$ unit masses of chlorine.[2] But unfortunately in cases where an element combines

---

[1] As a matter of fact, Wollaston's standard of equivalency was not unit mass of hydrogen, but ten unit masses of oxygen. This, however, does not in any way affect the line of argument in the text (see note 1, p. 95).

[2] It is to be observed that the equivalent of any element may be determined as well by investigating the mass of hydrogen which a given mass of the element *replaces*, as by finding the mass of hydrogen with which a given mass of the element *combines*. For every combination of an element with chlorine, bromine, &c., may be regarded as a substitution product of hydrochloric acid, hydrobromic acid, &c.

with both chlorine and hydrogen, the equivalent deduced from the chloride does not always coincide with that deduced from the hydride. Again some elements, notably carbon, form numerous compounds with hydrogen (methane, ethane, ethylene, &c., in the case of carbon), and each of these compounds would give a different value to the equivalent for carbon. Who was to decide which particular compound was to be selected for fixing the value in question? That an element should be possessed of several equivalents seemed a pure contradiction in terms.

Further complications arose when Gmelin and Gay-Lussac attempted to make equivalent the formulæ of all compound bodies, including the most important class of salts.[1] As long as the terms equivalent, or equivalent weight, are restricted to acids and bases among compound bodies, they have perfectly clear and definite meanings, and are still employed in this connection in modern volumetric analysis. But it is hard to appreciate the application of the terms to salts. However, this application was made in different ways by different chemists, with the result that instead of applying the equivalents of the elements deduced from hydrides, chlorides, &c., to transform the results of analysis of compounds into formulæ for these com-

---

[1] Here the idea of equivalency is somewhat changed. "$AgO.SO_3$, and NaO, SO_3, are equivalent, not because they have equal powers of displacement or combination with regard to any criterion, but because they are the results of the union of bodies in such proportions that the equal powers of combination of the constituent parts were satisfied in the act of combining."

pounds, the problem was often reversed, and the equivalents of the elements were deduced from presumed equivalences between salts.[1] The values obtained in this way did not in general agree with those directly obtained from analysis of hydrides, chlorides, &c. Thus both the values $4\frac{2}{3}$ and 14 for the equivalent of nitrogen found supporters; while the equivalent of phosphorus was either $10\frac{1}{3}$, 15·5, or 31.

It is almost impossible for us now to appreciate fully the reasoning of the chemists of this period. The confusion of ideas which prevailed,[2] the gratuitous

---

[1] See Wurtz, *The Atomic Theory*, p. 71, *et seq.*

[2] Here it may be well to distinguish between the three terms combining weight, equivalent, and atomic weight. The idea of combining weights was associated with, and early recognised as a necessary consequence of, the atomic theory (see note 1, p. 77). The atomic weights were either numerically equal to, or whole multiples of, certain characteristic numbers called combining weights that could be assigned to the elements from analyses of their compounds, quite independently of any assumptions or suppositions. Hence it would seem that the terms combining weight and equivalent mean essentially the same thing; they do. The only difference between them is a chronological one. The equivalent notation came after the atomic theory, and was intended to be independent of it. It was an attempt to replace what up to date had proved itself from a chemical standpoint an unsatisfactory theory. Combining weights had their names changed to equivalents when the atomic theory had been weighed and found wanting. The reason for this change in nomenclature is not far to seek. It is an epitomised history. The atomic weights of the Daltonian school were in many instances numerically equal to the combining weights assigned to the elements, and the two terms, though radically distinct, thus came to be used more or less synonymously. Wollaston, in proposing the name equivalent weights for the constants of his new system, was simply desirous of avoiding a term which had become, quite wrongly, more or less identified with a theory. The term equivalent was to connote fact, and fact alone.

The difference between the terms atomic weight on the one hand,

assumptions which flourished, and the reckless use of the analogic method, combine to render its history one of the most unsatisfactory and bewildering pieces of modern scientific literature. Owing to the numerical coincidences in many cases of the atomic weights and the equivalents of the elements, the two terms, though fundamentally so distinct in their connotations, came to be used indifferently and synonymously.[1]

It was Gerhardt and Laurent who, resuscitating the long-eclipsed hypothesis of Avogadro, led the way out of this confusion worse confounded. The equivalent system presented to the mind of Gerhardt great inconsistencies, which, in his opinion, could only be

---

and equivalent or combining weight on the other, is more fundamental than the purely chronological one distinguishing equivalent from combining weight. The term atomic weight implies a theory of the structure of matter, the other two terms do not. The atomic weight of chlorine is 35·5; its equivalent or combining weight is also 35·5. The former statement calls up the following mental picture. Chlorine gas is made up of a number of indivisible ultimate particles called atoms equal among themselves, and each $35\frac{1}{2}$ times as heavy as the similar ultimate particles of which a mass of hydrogen consists. Whereas the latter statement simply implies that a mass of chlorine weighs $35\frac{1}{2}$ times as much as the mass of hydrogen with which it chemically combines. Though all idea of an equivalent notation is now abandoned, the terms equivalent and combining weight are still used, but now synonymously to denote the smallest mass of an element that combines with unit mass of hydrogen, or $35\frac{1}{2}$ unit masses of chlorine, or with (approximately) 8 unit masses of oxygen. Indeed it will be shown further on that the equivalent, as thus defined, really determines the final value adopted for the atomic weight of an element.

[1] Even as late as the year 1834 we find a committee of the British Association for the Advancement of Science passing the following resolution : "We are of opinion that it would save much confusion if every chemist would always state explicitly the exact quantities which he intends to represent by his symbols."

removed by a partial return to the Berzelian system of atomic weights. The inconsistencies that especially appealed to Gerhardt were something of this nature. The equivalents of carbonic anhydride and water being represented as CO_2 and HO respectively ($C = 6, O = 8$), Gerhardt noticed that when these substances were produced in reactions involving organic bodies, they always appeared in such relative quantities that the accepted equivalents of the organic bodies could never be represented as giving rise to a *single* equivalent, but always to 2, 4, or more equivalents of the oxides of carbon and hydrogen.

Thus the interaction of acetic acid and sodium carbonate was represented as follows—

$$C_8H_8O_8 + 2\ NaCO_3 = C_8H_6O_8Na_2 + 2\ CO_2 + 2\ HO.$$

It seemed incongruous and unnatural that organic bodies should be so differently constituted from inorganic ones as to be incapable of furnishing *single* equivalents of water and carbonic anhydride when undergoing chemical transformations. Gerhardt therefore proposed a return to the Berzelian values $C = 12$, $O = 16$, $S = 32$. Adopting these values, the difference noted between organic and inorganic substances disappears. The formulæ for the equivalents of water, sodium carbonate, and acetic acid become now H_2O, Na_2CO_3, and $C_4H_8O_4$ respectively. The formula for the equivalent of carbonic anhydride still remains CO_2, the equivalent weight being, however, doubled. The equation representing the interaction between acetic acid and sodium carbonate now takes the form—

$$C_4H_8O_4 + Na_2CO_3 = C_4H_6Na_2O_4 + CO_2 + H_2O$$

involving only single equivalents of carbonic anhydride and water.

But there still remained in Gerhardt's opinion a want of uniformity in the equivalent notation. He noticed that the current formulæ representing the equivalent weights of compounds were not strictly comparable, in that quantities of bodies corresponding to their formulæ expressed in grams,[1] say, did not occupy the same volume when converted into the gaseous state under similar conditions. Thus, CO_2, $C_4H_{12}O_2$, and $C_4H_8O_4$ were the accepted formulæ for the equivalents of carbonic anhydride, alcohol, and acetic acid respectively.

	Equivalent Weight.	Volume in gaseous state (reduced to 0° and 760 mm.) of equivalent weight expressed in grams.
CO_2	44	22·4 litres
$C_4H_{12}O_2$ . . .	92	44·8 ,,
$C_4H_8O_4$ . . .	120	44·8 ,,

Now Gerhardt proposed to make such alterations as were necessary to reduce all formulæ to a common gaso-

---

[1] Of course the interpretation of formulæ in grams is quite arbitrary, but so long as we are ignorant of the absolute weight of the hydrogen atom we cannot fix the absolute weights of compound atoms. What we practically do is to make the conveniently simple and concrete supposition (which we know to be far from true) that the hydrogen atom weighs 1 gram, and the compound atoms of carbonic anhydride, alcohol, and ether, 44 grams, 92 grams, and 120 grams respectively.

metric standard. Those masses of substances were to
be equivalent which occupy the same volume in the
gaseous state, and the formulæ representing the compositions of these masses were to be the equivalent
formulæ. It is evident that this reform of Gerhardt's
is nothing else than a return to Avogadro's "hypothesis." Laurent clearly recognised this, and impressed
on Gerhardt the advisability of replacing the term
equivalent by molecule when it has reference to compounds ; and by molecule or atom, as the case may be
when it is used in connection with elements. Henceforth Laurent's nomenclature will be observed.

Although Gerhardt returned to the Berzelian atomic
weights, yet it is interesting to remark that considerations, not of an atomic structure of matter, but of an
equivalence, physical rather than chemical in its nature,
lay at the foundation of his reform, which ultimately
resulted in the complete abandonment of the equivalent
system so called.

The question as to which formulæ he was to alter
next presented itself to Gerhardt. Was he to double
the formula of carbonic anhydride so as to make the
smallest particle, conserving all the properties of the
substance in mass, occupy the same volume as did
the accepted molecule of alcohol. Or was he to halve
the formula of alcohol so as to make its molecule
occupy the same volume as the accepted molecule of
carbonic anhydride ? To settle this question he had to
determine what volume a standard molecule, such as
that of hydrogen, occupies. From the volumetric com-

position of hydrochloric acid gas, it follows that the molecules of hydrogen and chlorine must each contain an *even* number of atoms; two at the least. That the hydrogen molecule does not contain more than two atoms is proved by the fact that hydrochloric acid is a monobasic acid, and only under most exceptional conditions forms sodium and potassium salts containing hydrogen.[1]

Hence the molecular weight of hydrogen is 2, and, availing ourselves of the convenient supposition for theoretical purposes of note 1, p. 104, we may provisionally say that it weighs 2 grams. But it is a well-established constant that one litre of hydrogen under standard temperature and pressure weighs ·0896 ± grams. Therefore 2 grams of hydrogen will under these conditions occupy 22·4 ± litres.

Hence it follows that carbonic acid has the right formula, and that the formulæ of alcohol, acetic acid, and organic bodies generally must be halved.[2]

[1] Suppose for an instant that the molecule of hydrogen is tetratomic. Then the formula of hydrochloric acid would be H_2Cl_x, and it would necessarily be either a dibasic acid forming two sodium salts or a monobasic acid forming under all conditions salts of the type $M'HCl_x$, where M is a monovalent atom or radicle.

We have no good proof that the molecule of chlorine is not of higher atomicity than 2. However, chlorine is chemically very analogous to iodine, and it seems almost certain, from specific gravity determinations carried out on the vapour of this body through a wide temperature range, that its molecule does not contain more than 2 atoms.

[2] The doubled formulæ of organic bodies generally was due (1) to the fact that these formulæ were in many cases derived from analyses of silver derivatives, and Berzelius' value for the atomic weight of silver was twice too great; (2) to the fact that certain views (embraced under the title dualism) held at this period regarding the constitution

Gerhardt's proposed reform did not, however, meet with general acceptance till after the brilliant work of Williamson on the ethers. This research incontestably showed by purely chemical reasoning the necessity for halving the formula of alcohol in accordance with Gerhardt's views, *i.e.*, in accordance with the hypothesis of Avogadro.

Starting with ethyl alcohol, Williamson hoped to prepare therefrom an alcohol of higher molecular weight. To this end he first treated ethyl alcohol with potassium and then with ethyl iodide. Contrary to expectation, the product was not an alcohol at all, but ordinary ether $C_4H_{10}O$. This is inconsistent with the formula $C_4H_{12}O_2$ for ethyl alcohol; for if the latter body contains per molecule twice as much oxygen as ether, then the product of the above reactions ought to have contained twice as much oxygen as ether, because there has simply been replacement of H by C_2H_5. The only way out of this difficulty appeared to be the representation of alcohol by the halved formula C_2H_6O.

$$\text{Alcohol.} \qquad\qquad \text{Ether.}$$

$$\left.\begin{array}{l} C_2H_5 \\ H \end{array}\right\} O \qquad\qquad \left.\begin{array}{l} C_2H_5 \\ C_2H_5 \end{array}\right\} O$$

of bodies and the mechanism of chemical changes demanded the double formulæ for their expression. For details, see Meyer, *History of Chemistry*.

It follows that Gerhardt's halving of the molecular weights of acetic acid, alcohol, &c., necessitated the halving of the Berzelian values for the atomic weights of silver and the alkali metals. Unfortunately Gerhardt carried his views to an extreme, and, without good warrant, halved the atomic weights of twenty-three other elements, an error which was afterwards pointed out and rectified by Cannizzaro, who applied to the cases in question Dulong and Petit's law of specific heat.

True, a representation of this special formation of ether could be given in terms of the doubled formula, thus—

$$\left.\begin{array}{c} C_4H_{10}O \\ K_2O \end{array}\right\} + C_4H_{10}I_2 = C_4H_{10}O + C_4H_{10}O + 2KI$$

But according to this mode of representation, a mixture of methyl and ethyl ethers ought to result if methyl iodide were used instead of ethyl iodide.

$$\left.\begin{array}{c} C_4H_{10}O \\ K_2O \end{array}\right\} + C_2H_6I_2 = \underset{\text{(ethyl ether)}}{C_4H_{10}O} + \underset{\text{(methyl ether)}}{C_2H_6O} + 2\ KI.$$

As a matter of fact, however, a single ether—a so-called mixed ether—and not a mixture of two ethers is the result; and this can only be adequately represented in terms of the formula C_2H_6O for alcohol.

$$\left.\begin{array}{c} C_2H_5 \\ K \end{array}\right\}O + CH_3I = \underset{\text{(ethyl methyl ether)}}{\left.\begin{array}{c} C_2H_5 \\ CH_3 \end{array}\right\}O} + KI.$$

Accordingly the molecular weight of alcohol was halved, and thereupon the Gerhardt-Avogadro hypothesis began to grow in general favour, and to acquire that recognition which to-day classes it as the most important instrument for the determination of atomic weights.

Before demonstrating the use of the Gerhardt-Avogadro generalisation in the determination of atomic weights, we must first throw it into a more convenient equational form.

H	X	Y
1 gram	w grams	w' grams
n molecules	n molecules	n molecules

Let H represent a volume of hydrogen weighing 1 gram. Let X and Y represent equal volumes of other gases x and y, weighing respectively w and w' grams. Let there be n molecules of hydrogen present in the volume H; then, assuming the truth of Avogadro's generalisation, there will be n molecules of x in the equal volume X and n of y in the equal volume Y.

Now the specific gravity[1] of x (referred to hydrogen as standard) is obviously w.

And the specific gravity of y is w'.

Further, the weight of one molecule of x must be $\frac{w}{n}$ grams, and the weight of one molecule of y must be $\frac{w'}{n}$. Hence—

$$\frac{\text{Molecular weight of } x}{\text{Molecular weight of } y} = \frac{\frac{w}{n}}{\frac{w'}{n}} = \frac{w}{w'} = \frac{\text{Specific gravity of } x.}{\text{Specific gravity of } y.}$$

In other words, Avogadro's generalisation may be re-

---

[1] In chemistry, the term " vapour-density " is generally used when "specific gravity of vapour " is really meant. Density is simply mass per unit volume ; it is of two dimensions. Specific gravity is the ratio of the density of a body to that of some standard substance. The number expressing its value for any given substance is a pure number and has therefore no dimensions.

stated in the form: the molecular weights of gases vary directly as their specific gravities.[1]

Since x and y are any two gases whatever, let y represent hydrogen.

Then—

$$\frac{\text{Molecular weight of } x}{\text{Molecular weight of hydrogen}} = \frac{\text{Specific gravity of } x.}{\text{Specific gravity of hydrogen.}}$$

$$\therefore \frac{\text{Molecular weight of } x}{2} = \frac{\text{Specific gravity of } x.}{1}$$

or, molecular weight of $x = 2 \times$ specific gravity of x.

If, as is frequently the case, air be taken as the standard of specific gravity, the above equation becomes—

$$\text{Molecular weight of } x = 2 \times 14\cdot44 \times \text{specific gravity of } x,$$
$$= 28\cdot88 \times \text{specific gravity of } x,$$

air being, bulk for bulk, approximately 14·44 times heavier than hydrogen.

Having now thrown Avogadro's generalisation into a numerical and equational form, we proceed to apply it

---

[1] Avogadro's law admits of still another mode of expression. The gaseous laws of Boyle and Charles are succinctly summed up in the equation

$$pv = \text{RT}.$$

Where p is the pressure, v the volume, and T the absolute temperature of a given mass of any gas, R is a constant depending (1) on the mass of the gas taken, and (2) on the nature of the gas. However, if we agree to apply the equation in all cases to masses of gases equal to their Avogadrean molecular weights interpreted in grams, then R no longer varies from gas to gas, but has the constant approximate value, 84,700 (the pressure being expressed in gravitation units—grams per square centimetre).

to the determination of an atomic weight, say, to the determination of the atomic weight of oxygen.

I.	II.	III.	IV.	V.	
Carbon monoxide .	42·86 C : 57·14 O	14	28	12 C	: 16 O
Carbon dioxide .	27·27 C : 72·73 O	22	44	12 C	: 32 O
Osmic oxide .	74·93 Os : 25·07 O	127·6	255·2	191·3 Os	: 64 O
Water . .	11·11 H : 88·89 O	9	18	2 H	: 16 O
Arsenious anhydride	75.78 As : 24·22 O	198	396	300·4 As	: 96 O
Nitric oxide .	46·67 N : 53·33 O	15	30	14 N	: 16 O
Sulphuric anhydride	40·00 Z : 60·00 O	40	80	32 S	: 48 O
&c.	&c.	&c.	&c.	&c.	

Having selected several—the larger the number the better—gaseous or gasifiable compounds of oxygen (column i.), whose percentage compositions (column ii.) are known to a fair degree of accuracy, we find their approximate specific gravities in the gaseous state [1] (column iii.). From these values we deduce the approximate molecular weights [2] (column iv.) by means of one

[1] For practical details see Muir and Carnegie, *Practical Chemistry*, p. 121.

[2] The definitions of molecular weight given in many of the text-books are exceedingly vague, *e.g.* :—

"The molecular weights are the weights of two volumes, for molecules occupy two volumes if an atom of hydrogen occupies one."

"The molecular weight of a gaseous element or compound is a number which tells the weight of two volumes of the gas, that is, the weight of that volume of the gas which is equal to the volume occupied by two parts by weight of hydrogen."

In the following definition, the attempt is made to avoid the indefinite terms "volumes" and "parts by weight."

The molecular weight of any substance is that weight thereof (expressed in terms of any unit whatever) which in the gaseous state occupies the same volume as do two unit weights of hydrogen, the same conditions of temperature and pressure and the same weight unit being observed in both cases.

or the other of the equations just developed. Then we re-state the percentage compositions in such a way as to give the molecular compositions (column v.). The ratio $12 : 16$ (column v.) is just the same thing as the ratio $42\cdot86 : 57\cdot14$ (column ii.), but $12 + 16 = 28$ the molecular weight of carbon monoxide. And so on for all the other bodies mentioned.

Now these molecular magnitudes have been deduced from equations which are based on the supposition that the smallest mass of hydrogen ever found in any molecule (*i.e.*, the atom of hydrogen), is numerically represented by unity; and a glance at column v. convinces us that the smallest mass of oxygen found in the molecule of any compound considered is, in terms of the same mass unit, represented numerically and approximately by 16. But the least amount of oxygen that can by theory exist in a molecule is an atom. Therefore the approximate atomic weight of oxygen is 16.[1] The method of procedure in the case of any other element is exactly similar.

It still remains to show how the approximate atomic weight assists us to a knowledge of the true atomic weight, which now follows upon a very accurate determination of the equivalent of the element, *i.e.*, the

---

[1] We cannot positively and finally assert that the atomic weight of oxygen is 16. We can only say that the probability of the approximate value O = 16 is almost infinitely great. A new substance *might* be discovered whose molecule contained only $8\pm$ unit masses of oxygen (H atom = 1 unit mass); and there are chemists who, desirous of emphasising this vanishing possibility, speak of the *maximum* atomic weight of oxygen being approximately equal to 16.

smallest mass of the element which combines with unit mass of hydrogen, or with that mass of oxygen which itself combines with unit mass of hydrogen.

In the case of oxygen the equivalent is not yet definitely agreed upon (see note 1, p. 95). As a probable result from all the recent elaborate determinations of this most important constant, Ostwald (*Lehrbuch der Allgcmeinen Chemie*, p. 48) gives the value 7·974 [H = 1]. Suppose W_h grams of hydrogen combine with W_x grams of an element X, atomic weight Z, to form a hydride whose molecular formula is $H_m X_n$. Then $\dfrac{W_x}{W_h}$ is the equivalent of the element X, and the ratio

$$\frac{Z \times n}{m} = \frac{W_x}{W_h}$$

$$\text{or } Z = \text{equivalent number} \times \frac{m}{n}$$

necessarily holds. In other words, the atomic weight of any element is numerically equal to its equivalent number multiplied by a fraction involving, as a rule, only the lower digits. In general, m is a whole multiple of n, so that the value of the atomic weight becomes a whole multiple of the equivalent number, or

$$Z = r \times \text{equivalent number,}$$

where $r = 1, 2, 3, \&c.$

In the case of oxygen, then, it follows that the atomic weight may be either

$$7 \cdot 974$$
$$\text{or } 15 \cdot 948 = (7 \cdot 974 \times 2)$$
$$\text{or } 23 \cdot 922 = (7 \cdot 974 \times 3) \text{ &c.}$$

The approximate value 16, already arrived at by the application of Avogadro's generalisation, enables us to select without any hesitation from these possibilities the value 15·948 as the interim atomic weight of oxygen.

Until Raoult's recent work, the constancy in the volumes occupied by gaseous molecules (Avogadro's generalisation) was the only known example of a colligative property.[1]   Raoult established the fact that

[1] Following Ostwald, we may divide all chemico-physical properties into three classes ; additive, constitutive, and colligative. Weight is the one and only true example of an additive property, for the weight of a compound is exactly equal to the weights of the elements which have reacted to form the compound. The majority of chemico-physical properties which have been quantitatively investigated, belong to the class of constitutive properties. Let A B D and A B C represent two compounds of distinct types, and X', X'' the respective values for these bodies of some constitutive property, such as molecular volume $\left(= \dfrac{\text{molecular weight}}{\text{specific gravity}} \right)$. Let a, b, c and d represent the atomic volumes $\left(= \dfrac{\text{atomic weight}}{\text{specific gravity}} \right)$ of A, B, C and D, respectively, in the free state. Then X' and X'' are not necessarily equal to $a + b + d$ and $a + b + c$, respectively ; for A B and D in compounds of the type A B D may have volumes a' b' and d', while in compounds of the type A B C, A and B may have still other values a'' and b''.   Hence

$$X' = a' + b' + d'$$
$$X'' = a'' + b'' + c''.$$

It will be seen, however, that constitutive properties involve an additive

dilute solutions of indifferent bodies also exhibit well-marked colligative properties.

A given quantity of solvent has its freezing point, or, if volatile, its vapour pressure for a given temperature, lowered a constant amount in each case, by the solution therein of different substances, in quantities proportional to their molecular weight. Just as the colligative property expressed in Avogadro's generalisation gave rise to a method for molecular weight determination (and thence for atomic weight valuation), so Raoult's results have in the hands of Beckman and others led to new practical methods of great utility for the determination of molecular weights. These new methods are all the more valuable in that they can be applied to bodies which cannot be conveniently gasified, or cannot be gasified without decomposition, i.e., to bodies which, owing either to refractoriness or instability, fall outside the pale of Avogadro's generalisation.

Since Van't Hoff showed that Raoult's empirical laws can be derived by thermodynamical reasoning one from the other, and both from a certain theory of solution suggested by the phenomena of osmotic pressure, they have been very widely and generally applied to the determinations of molecular weights in cases unsuited to the gaseous specific gravity method.

element ; and indeed, as soon as we know the values of any property peculiar to the atoms *in a given class of compounds*, the evaluation of constitutive properties for bodies of that class is a matter of pure addition. Avogadro's generalisation is the expression of a colligative property. No matter what the composition or complexity of a gaseous molecule may be, it will occupy under given conditions a given determined space.

Of the two methods due to Raoult, that founded on the lowering of freezing point seems the most practical and popular; it has therefore received a special name, being known as the cryoscopic method. For the details of the practice and theory of these methods the reader is referred to Ostwald's *Solutions*, translated from the *Lehrbuch der Allgemeinen Chemie*.

It cannot be too strongly emphasised that chemists, in applying to bodies in the solid and liquid states the constants determined for the molecules of these bodies in the gaseous state, do not thereby intentionally commit themselves to any theory of the molecular structure of solids and liquids. Chemical changes find adequate interpretation in terms of the formulæ derived from a study of gases; indeed, it seems as if the complex molecules of solids and liquids, under the necessary conditions for chemical change, and prior to such change, break down into simpler molecules, comparable, if not identical in magnitude, with the gaseous molecules.[1] In other words, it happens that the gaseous molecule remains the chemically reacting unit in the solid or liquid state, although the physical molecules (*i.e.*, the smallest masses of solids and liquids which exhibit all the properties, *physical* and chemical, of the substances

---

[1] Seeing that the unit of chemical activity of a solid or liquid is not necessarily the physico-chemical molecule of that substance, but may be only a dissociation product thereof, comparable in magnitude with the true gaseous molecule, there are those who advocate the restriction of the terms molecule and molecular weight to purely gaseous phenomena, using the non-committal terms, reacting unit and reacting weight, in describing the interactions of solids and liquids.

peculiar to their solid or liquid states), are almost certainly polymers of the gaseous molecules.

If MN be the formula of the gaseous molecule, then $(M_x N_x)$ and $(M_y N_y)$ represent the physical molecules of the same substance in the liquid and solid states respectively; x and y being whole numbers, and y in all probability greater than x.

It may be that the time is coming when we shall express more in our equations than we do at present, this increased information involving the true molecular magnitudes of the solids and liquids (or dissolved substances) interacting, *i.e.*, the absolute values of x and y.

That this polymerisation, or condensation of the simpler into more complex molecules, does often take place as a substance passes from the gaseous into the liquid state is certain, if we are justified in the universal application to vapours of Avogadro's generalisation. With many substances it makes itself evident even before the liquid state is reached. At high temperatures and low pressures gaseous nitric peroxide has the triatomic molecule NO_2; at lower temperatures and higher pressures these simpler molecules combine in pairs to form hexatomic molecules N_2O_4. Again, acetic acid vapour at high temperatures is made up of molecules $C_2H_4O_2$, while at temperatures near its condensing point the molecules have the more complex formula $C_4H_8O_4$. Indeed in many cases the specific gravities of the vapours of substances gradually increase as the temperature of the boiling-point is approached, indicating a gradual increase in complexity of the molecules.[1]

Now the gaseous and liquid states are really continuous, despite the apparent discontinuity which appears in the majority of cases under ordinary conditions of temperature and pressure. Hence we are led to the conclusion that the molecular condensation which appears in some cases as the liquid state is approached (but in other cases is so delayed as not to manifest itself until the upper limiting temperature of the liquid state has been actually passed) will continue to take place in increasing degree as the temperature is lowered after the liquid state is reached.

That this conclusion is just has been verified in

---

[1] The variability in the values of specific gravity determinations of vapours at different temperatures is well manifested in the case of the chlorides of aluminium and iron, and in this connection has caused much discussion. Grünewald and Meyer hold that the specific gravity of ferric chloride does not become approximately constant for any considerable temperature interval until a temperature of over 700° has been reached, when the vapour consists of molecules having the composition $FeCl_3$. On the other hand, Nilson and Pettersson maintain that the specific gravity of the vapour is constant for quite a large temperature interval, with 321° as its lower limit. Their experiments led to the molecular formula Fe_2Cl_6. Again, while Friedel and Crafts upheld the formula Al_2Cl_6 for aluminic chloride, Nilson and Pettersson are led by their experiments to support the formula $AlCl_3$. It seems necessary to admit that in both cases molecules of the types M_2Cl_6 and MCl_3 exist ; an admission which surrounds the expression "molecular weight of a gas" with indefiniteness. For all purely equational purposes it matters little which formula we use; but the true molecular formula, if indeed we can speak of a true molecular formula in these cases, is a point of great moment regarded from the standpoint of the valency theory (see Chapter VI.). Obviously the whole discussion turns on the difficulty of proving the existence and nature of the decompositions and dissociations taking place in a gas at high temperatures.

several cases by the application of Raoult's cryoscopic method of molecular weight determination.[1] Yet it must be remembered that the cryoscopic method can only be applied to fairly dilute solutions, and it is almost certain that in such solutions the molecule of a dissolved solid or liquid is smaller than it would be in the homogeneous solid or liquid state. The mere act of solution often dissociates the molecules of the solid or liquid, to such an extent that the dissociated molecules are of the same magnitude as the gaseous molecules; sometimes, however, the dissociation does not proceed so far, the extent of the dissociation being dependent on the nature of the indifferent solvent used. Acetic acid, used as a solvent, seems to have a great dissociating tendency; hydrocarbon solvents, on the other hand, manifest this property in a much less pronounced degree.[2] In short the cryoscopic method can only

[1] Measurements of other physical quantities also point to the existence of complex liquid molecules. Eötvös finds it impossible to explain certain observations of his on surface tension, unless it be admitted that molecules of liquids are from case to case more complex in varying degree than the molecules of the same substances in the gaseous state. Recently Ramsay and Shields have come to similar general conclusions. See *Chemical Society's Journal*, lxiii. p. 1089.

[2] Because certain liquids (*e.g.*, acetic ether) possess in *all* solvents the same molecular weight as they do in the gaseous state, it has been suggested that the gaseous molecules of such bodies do not polymerise when liquefaction takes place. This seems quite a misleading suggestion. It is *not* probable that the dissociation tendency is a function only of the solvent; the greater or less stability of the molecular complex of the dissolved substance itself must also be a factor. And surely it is not improbable that in some bodies the molecular complexes peculiar to the liquid state may be held together so loosely that they break down even under the influence of comparatively inert hydrocarbon solvents.

furnish us with a minimum value for the molecular weight of a solid or liquid body.

Of the true molecular weight of solids nothing is known. It is supposed that the polymerisation of the molecules in the solid state is greater than it is in the liquid state—that the molecule of ice is heavier than the molecule of water. Perhaps one and the same solid substance can have different weighted molecules according to the conditions of its formation and existence, and with this difference may, perhaps, be associated the phenomena of allotropism (or physical isomerism) and polymorphism (see p. 93, and Chap. VI.).

CHAPTER V.

THE CLASSIFICATION OF COMPOUNDS. ACIDS, BASES, SALTS.

BEFORE dealing with the more important classes of compounds, it is advisable to preface a few remarks on a classification of the elements which, in spite of indefiniteness and artificiality, still enjoys a greater or less degree of popularity.

At one time it was believed that chemical action is essentially electrical in its nature, and that every chemical change is the result of the play of the stronger or weaker attractions of oppositely electrified atoms or groups of atoms. Berzelius especially developed this view of the electrical nature of chemical affinity. He believed each atom to be electrically bipolar—each atom to have a definite charge of positive, and a definite charge of negative, electricity.[1] Since these charges were in general supposed to be unequal in amount, the charge present in greater amount gave a more or less positive or negative character to the atom as a whole. In accordance with the results of experiments on the behaviour of several compounds when

---

[1] If Berzelius had expressed himself in terms of electric potential, the necessity for assuming a bipolar electrical distribution in the atom would have been obviated.

electrolysed (*i.e.*, when decomposed by the electric current), Berzelius arranged the elements in an electro-chemical series such that the element with the most negative atoms stood at the head, and the element with the most positive atoms at the bottom, of the series, any intermediate element being more negative than those below it in the series, and more positive than those above it.

It was soon recognised that the elements towards the electro-negative end of the series thus constructed had properties in common which were different from those common to the elements towards the electro-positive end of the series. The electro-negative elements are, as a rule, of comparatively low specific gravity, bad conductors of heat and electricity, and more or less transparent. Their oxides are for the most part readily soluble in water, and produce solutions with a sour taste and great solvent and corrosive powers. The elements themselves show an aptitude for combining with hydrogen to produce hydrides and decompose water, combining with its hydrogen and setting free its oxygen.

The electro-positive elements, on the other hand, have as a class relatively high specific gravities, and possess a peculiar lustre. They are good conductors of heat and electricity, and are translucent only when reduced to very thin layers. Their oxides are for the most part insoluble in water, but they have the power of reacting with the solutions of the oxides of the electro-negative elements so as to destroy all the characteristic properties of the latter. Further, the electro-positive elements do not readily form compounds with hydrogen, and decompose water, combining with its oxygen and liberating the hydrogen.

It seemed good therefore to divide the elements into two classes.—

    (1) The electro-negative elements or the non-metals.[1]
    (2) The electro-positive elements or the metals.

If this classification enabled us to make anything approaching to a satisfactory definition of acids, bases, and salts, it might despite its indefiniteness [2] be retained. But we shall show further on that it does not possess this much to be desired merit, and may also remark here that the only satisfactory classification of the elements—that afforded by the periodic law—does not favour any such simple and fundamental division as is embraced in the terms metal and non-metal.

Before passing on to acids, bases, and salts, it is also

[1] By some the word "metalloid" is used in preference to "non-metal." In view of the significance of the termination —oid as generally used, this preference is scarcely justified.

[2] This indefiniteness is a necessary accompaniment of all classifications which are not based on *one* definite criterion. A number of properties are currently regarded as metallic, a number of other properties as non-metallic. These properties in each case are not considered of equal value *inter se* in deciding as to the metallic or non-metallic character of the element possessing them, but yet there exists no authoritative conventional scale of values. An element as a rule possesses properties falling under each category, and the non-metallic properties have to be weighed against the metallic ones in the unadjusted balances of private opinion. In the absence of all convention it is impossible to give a final and indisputable answer to the question, "Is tellurium a metal or a non-metal?"

necessary to say a few words on a thoroughly arbitrary classification of compounds which is in vogue. All compounds, by convention, fall under one or other of the two headings—organic and inorganic; and hand in hand with this classification goes a division of the science at large into two branches—organic and inorganic chemistry. But "carbon compounds" and "the chemistry of carbon compounds" are undoubtedly much more appropriate titles than "organic compounds" and "organic chemistry" respectively. For comparatively very few of the bodies treated in this main subdivision of compounds are, as the latter titles seem to imply, products of organisms, i.e., of life; while, on the other hand, some substances included among inorganic compounds *are* products of vital metabolism.

Since Lémery's time (1645–1715) up till the beginning of this century, chemists were possessed of the idea that the products of organisms—most of which are compounds of carbon—could not be made artificially, but that their production absolutely demanded the intermediacy of life—the operation of the so-called vital force. "Operations of chymistry fall short of vital force; no chymist can make milk or blood of grass." Further, it was the custom to ascribe the variable results obtained in analyses of animal and vegetable products to the fact that such bodies did not conform to the fundamental law of chemistry (law of fixity of composition), and not to the difficulties attendant on the analytical methods and the purification of complex carbon compounds. Hence a very sharp distinction was made between organic and inorganic or mineral compounds.

But the distinction is now recognised as existing for convenience only.[1] Exactly the same laws hold in the case of carbon compounds as in the realm of mineral or inorganic chemistry, and many of those carbon compounds which are products of the life of organisms have been synthesised independently of the living laboratory, *i.e.*, of vital force so called. Thus there exists no better reason than convenience to be adduced for this primary division of compounds, and no valid excuse exists for retaining the effete and misleading term "organic compounds" in preference to the more fitting title "carbon compounds." The compounds of nitrogen are increasing so rapidly, that before long it may be advisable to study them as a class apart, just as has been the case with carbon compounds.

In discussing acids, bases, and salts, it will be only too apparent that we are unable to give satisfactory definitions of these classes of compounds, and farther, that any of the attempts at definition that may be advanced are not independent one of another. In other words, we cannot attempt to define one of these classes of compounds without implying or assuming the definitions of the other classes. In physics all definitions can be expressed quite independently of one

---

[1] A sign of the extremely arbitrary nature of this classification is seen in the fact that the oxides of carbon, together with the bodies they form by direct combination, are included in inorganic chemistry.

another in terms of powers of three fundamental units, but in chemistry we have no approach to such definition.

Acids.—Probably the first acid known to the ancients, certainly the one best known to them, was vinegar. Regarding the solvent powers of this comparatively feeble acid, most exaggerated notions were entertained ; witness the belief that Hannibal therewith etched a passage over the Alps for his army. The alchemists, however, knew most of the mineral acids. Geber (eighth century) describes the preparation of *aqua fortis* (nitric acid) and also of *aqua regia*, which he obtained by dissolving sal ammoniac in *aqua fortis*. Geber also knew of oil of vitriol. The iatro-chemist Valentine (beginning of the sixteenth century) first made *spiritus salis* (hydrochloric acid), and showed that *aqua regia* can be made by mixing this new *spiritus* with *aqua fortis*. Later iatro-chemists, Libavius and Glauber, contributed much to an increased knowledge and more extended use of the acids by introducing improved methods of preparation.

But as yet no attempt at the classification of bodies by their properties had been made, and the generic term acid had not arisen.

Boyle was the first to group together into one class all substances which have the following properties : (1) a sour taste, (2) a great solvent action, (3) the power of precipitating sulphur from alkaline solutions of this substance, (4) a reddening action on many vegetable blues such as litmus, (5) the power of acting on wood ashes to produce substances without either the astringent solvent properties of acids or the soapy cleansing properties of solutions of wood ashes. Bodies possessing these properties were called collectively acids. Even down to the beginning of the eighteenth century most chemists were satisfied with the explanation that all substances possessing these properties did so in virtue of a greater or less amount of a common constituent which was called the "primordial acid." Such a vague explanation did not however satisfy Lavoisier, who renounced all alchemistic fancies of "primordial acids" and "principles of acidity." He thought, as we have seen, that oxygen was the acid producer, and that therefore every acid owes its acidity in some way to the oxygen which it necessarily contains. It must be here emphasised that Lavoisier did not regard the elements of water, and therefore hydrogen, as necessary constituents of acids. His acids were mere disseminations in water of our acid anhydrides (or acidic oxides)—binary compounds of some non-metal with oxygen. Thus to Lavoisier sulphuric acid was SO_3, and not H_2O, SO_3, or H_2SO_4.[1]

Up to the year 1787 all the substances known to

---

[1] It is now believed that after their solution in water, the acid anhydrides as such lose their identity ; a new class of bodies called hydroxides being formed. These bodies contain as proximate parts of their molecules one or more hydroxyl (OH) radicles. Thus when SO_3 reacts with H_2O, a rearrangement of atoms is believed to accompany combination, so that the resulting compound is represented as $SO_2(OH)_2$ (sulphuryl di-hydroxide), and not as SO_3, H_2O (hydrated sulphuric oxide). Similar remarks apply to the solutions of the oxides of certain metals known as the alkaline oxides, and the alkaline-earth oxides. Thus potassium oxide K_2O reacts with water in the process of dissolving in it to form potassium hydroxide $K(OH)$.

have the properties of acids as laid down by Boyle were products of the interaction of water with the oxides of non-metals; hence the oxides of the non-metals were called acid (acidic) oxides. In the year just mentioned Berthollet took up the investigation of prussic acid (discovered some three or four years previously by Scheele), and came to the conclusion that it had true acid properties but yet was entirely free from oxygen. In 1796 he came to the same conclusion for sulphuretted hydrogen, or sulphydric acid, as it may be called. Lavoisier's reputation was, however, more mighty than Berthollet's facts, and as regards acids, matters remained *in statu quo* until Davy in 1810 clearly proved that hydrochloric acid does not contain any oxygen, but is a compound of hydrogen with an element chlorine. Further, in 1813, hydriodic acid was shown to be altogether free from oxygen, and it was remarked that iodic anhydride shows no acid properties until it is dissolved in water, *i.e.*, until the element hydrogen is introduced. Davy then began to recognise that there is no one particular acid-forming element. He noticed that hydrogen is a universal constituent of acids, but he did not rush over to the converse and say that all hydrogenised bodies are necessarily acids. On the contrary, he regarded the existence of acid properties as depending chiefly on the other elements combined with the hydrogen, and not on the hydrogen itself. The result of Davy's work was that acids came to be classed as hydracids (acids free from oxygen) and oxyacids (acids formed from acidic oxides). Shortly afterwards Liebig came to the same conclusions as Davy, and defined acids as particular compounds of hydrogen in which the latter can be replaced by metals.

The growth of our knowledge since Davy's time has not brought us a satisfactory expression of any generalisation as to the particular compounds of hydrogen enjoying this property. We are in fact bound to admit that an acid belongs to that category of things which are perfectly conceivable but indefinable. With Boyle and Liebig we can enumerate the properties which we agree shall be connoted by the term acid—we can agree as to what are to be regarded as acidic functions, but that is all. The difficulty remains that several bodies possessing some, or may be all, of these properties are, for other reasons, not regarded as acids. We have no one property which we can use as an absolute criterion of acidity. Thus the salt copper sulphate turns blue litmus red, and causes an effervescence of carbon dioxide when mixed with a soluble carbonate. Bisulphate of sodium turns blue litmus red, causes effervescence with soluble carbonate, and contains hydrogen which is replaceable by metals; yet on account of the method of its formation it is not regarded as an acid. It seems hopeless then to attempt any definition of the term acid in terms of chemical properties.

If we attempt a definition from the point of view of composition rather than properties, we meet with

I

equally great difficulties. As Davy and Liebig saw, all bodies regarded as acids contain hydrogen, but all hydrogenised bodies are not acids. Evidently the hydrogen must be combined with certain elements, and in a definite manner, to produce an acid. Let us try to generalise in this direction. The oxides of the non-metals react with water to produce hydroxides which are called acids (see note, p. 127). Therefore we might attempt to define acids as the hydroxides of the non-metals [Cl(OH)] or of non-metallic groups [$SO_2(OH)_2$]. But the bodies HCl, HBr, HI, &c., are called acids. Hence we must enlarge our definition to include these hydracids. Acids are the hydrides and hydroxides of non-metallic elements. But this is not a successful definition; for nitrogen and phosphorus are generally regarded as non-metals by those who employ the latter term, yet their hydrides are not acidic.[1] Again, the bodies H_2PtCl_6 and $H_4FeC_6N_6$ are acids, but neither $PtCl_6$ nor FeC_6N_6 can be called groups of non-metallic elements. It seems hopeless then, even had the title non-metal a strict connotation, to attempt to satisfactorily define an acid in terms of its qualitative composition. We cannot define an inorganic acid in terms of its properties alone, nor in terms of its composition alone, nor in terms of both together; composition, properties, and synthetical history must all be taken into account in deciding a body's claim to rank as an acid.

---

[1] Ammonia nevertheless contains hydrogen directly replaceable by the metals K, Na.

Acids are classified according to their strengths or affinities, and also according to their basicities. The former classification will be taken up in the chapter on chemical equilibrium. The latter classification is due to Liebig, who distinguished acids as mono-, di-, tri-, tetra-basic, &c., according as they possessed one, two, three, four, or more atoms of hydrogen per molecule replaceable by metals. It is not always true that the replaceable hydrogen atoms are coequal in number with the total hydrogen atoms present in the molecule. In cases where this equality exists the basicity of the acids can at once be determined from their formulæ; but the equality must have been proved by experiment to exist. The experimental investigation of basicity takes the form of a determination of the maximum number of stable potassium or sodium salts the acids can form.[1]

Bases.—From early times it had been noticed that the extracts of the ashes of burnt plants possessed certain characteristic properties. For example, these extracts have great cleansing power, apparently dissolving grease and fat; they are soapy to the touch, and have the property of restoring to their normal colours vege-table blues which have been reddened by acids. The active principle of these extracts was called an alkali, this title first appearing in Geber's writings, and liter-ally meaning *the ash*.[2] Soon it was discovered that a

[1] For practical details see Muir and Carnegie, *Practical Chemistry*, p. 38 *et seq.*

[2] The active principle of wood ashes is in reality potassium carbonate. This is no longer classed as an alkali; it is a salt with alkaline pro-perties.

solution of the highly volatile ammonia or hartshorn also possessed the above properties. Hence a distinction was made between fixed and volatile alkalis. It was Stahl who first noticed that, in addition to the above enumerated properties, alkalis have the power of reacting with acids to produce indifferent substances. It was as if the properties of acids and alkalis were diametrically opposed, and in combination mutual neutralisation took place, just as would be the case with two equal and opposite magnetic poles—a sort of cancelling out of positive and negative.

So far the substances which had the power of thus neutralising acids were all soluble in water. Sometime later it was found that certain almost insoluble substances which are very fixed in the fire, *i.e.*, do not melt nor change in any way, have the same property. These were called the earths, and Rouelle in 1744 combined alkalis and earths into the one class *base*, defining a base as any substance which combines with and neutralises an acid.

It is almost impossible to give a concise and exact definition of base as that term is now applied. Speaking loosely, the oxides of the metals, and the hydroxides they form when they are made to combine directly or indirectly with water, are regarded as bases; hence metallic oxides are often spoken of as basic oxides. May we not then define a base as the oxide or hydroxide of a metal? Such a definition does not include hydrides of certain non-metals (ammonia NH_3, phosphine PH_3, &c.), which are established as bases;

and, further, it does not indicate that the higher oxides of some metals have acidic functions.   We are equally unsuccessful when we try to frame a definition in terms of properties.   All bases do not react with acids so as to neutralise the latter, *i.e.*, completely destroy their characteristic properties, as Rouelle, to whom only the strongest bases were known, believed.   The solution obtained when sulphuric acid is treated with an excess of the base copper oxide still possesses acid characters.   Again, while some bases act on vegetable colouring matters, others do not.   Nor can bases be said to be bodies which in reacting with acids cause a replacement of the hydrogen of the acids by metals, for the bases N_2H_4, NH_3, PH_3, &c., do not contain metals, and the products of their interaction with acids are of the nature of additive, not substituted, compounds. True, the definition, " A base is a substance which reacts with acids to produce a salt and water at most," holds good, but it is a definition in terms of acids which cannot be defined, and involving a new class of bodies, salts which themselves require definition.   Hence we must conclude that while it is possible to lay down what we mean by basic characters or functions, it is impossible to define bases; for many bodies which are not regarded as bases exhibit basic characters, and many bases exhibit only a few of what are all admittedly basic functions.

A rough classification of the more important inorganic bases follows.   The table clearly shows the true relation of the terms alkali and base.   In some elementary text-books the terms are used as if they were synonymous. This is not the case. An alkali is a particular kind of hydroxide which is a particular kind of base. Every alkali is a base, but every base is not an alkali.

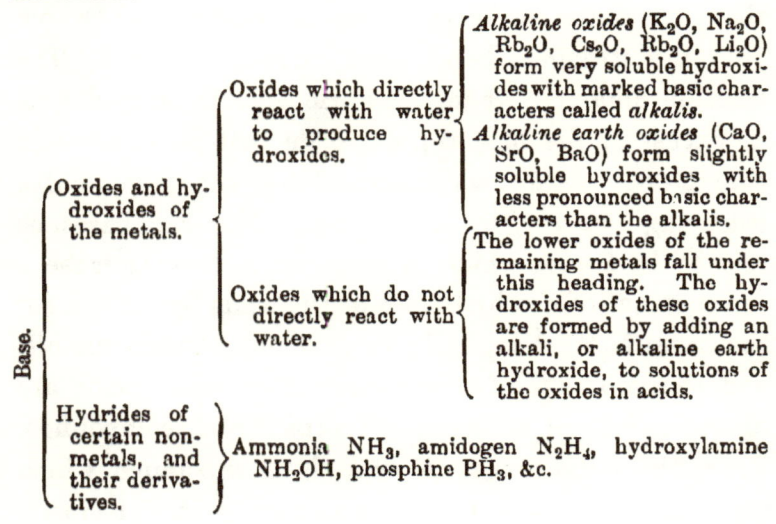

Base.
- Oxides and hydroxides of the metals.
  - Oxides which directly react with water to produce hydroxides.
    - *Alkaline oxides* (K_2O, Na_2O, Rb_2O, Cs_2O, Rb_2O, Li_2O) form very soluble hydroxides with marked basic characters called *alkalis*.
    - *Alkaline earth oxides* (CaO, SrO, BaO) form slightly soluble hydroxides with less pronounced basic characters than the alkalis.
  - Oxides which do not directly react with water.
    - The lower oxides of the remaining metals fall under this heading. The hydroxides of these oxides are formed by adding an alkali, or alkaline earth hydroxide, to solutions of the oxides in acids.
- Hydrides of certain non-metals, and their derivatives.
  - Ammonia NH_3, amidogen N_2H_4, hydroxylamine NH_2OH, phosphine PH_3, &c.

However convenient classifications of oxides into acidic, basic, intermediate, and indifferent varieties may be for an elementary presentment of chemistry, a glance from the vantage-ground of facts which are not usually referred to in elementary courses, but which are none the less facts on that account, shows such classifications to be imperfect and arbitrary to a degree. New reactions and new substances are continually being discovered, which testify to the fact that oxides generally are capable of exhibiting under one set of conditions what are currently accepted as acid characters; under another set of conditions, what are currently accepted as basic characters.   For instance, chromium trioxide is usually classed as an acidic oxide, because it reacts readily with the alkalis, partially destroying the characteristic properties of the latter. But it also reacts with sulphuric acid, destroying its characteristic properties, and surely this is a basic function.   True, the salt formed in this latter case does not *correspond to* CrO_3, but it has not yet been authoritatively laid down that such correspondence is an essential feature of acids, or bases as the case may be.

It seems, however, to be tacitly assumed that an oxide has no claim to rank as acidic or basic until it can be proved to produce *corresponding salts*. Thus, true basic rank was denied to PbO_2 until the recent preparation of the corresponding salts $PbCl_4$, $Pb(C_2H_3O_2)_4$, &c.   In spite of its reaction with hydrochloric acid, it was regarded as a feeble acidic oxide destitute of true basic properties.   Now in virtue of these new salts it ranks as an intermediate oxide.[1]

---

[1] When an acid and base interact to produce a salt only, or a mixture of salt and water only, the salt is said to correspond to both the acidic oxide of the acid and the basic oxide of the base.

$$BaO + H_2SO_4 = BaSO_4 + H_2O$$
$$2 BaO_2 + 2 H_2SO_4 = 2 BaSO_4 + 2 H_2O + O_2.$$

According to definition, $BaSO_4$ corresponds to the oxides SO_3 and BaO, but it does not correspond to the oxide BaO_2.   If other factors (*e.g.*, oxygen from the air) than the acid and base be involved, the resulting salt is not regarded as corresponding to one or other of the oxides involved.   Thus potassium manganate K_2MnO_4 (regarded as K_2O MnO_3) does not correspond to the oxide MnO_2 from which it is prepared.

Even if the restriction of correspondence (as defined in the note) were imposed, the difficulties attendant on the classification of oxides would not be greatly lessened. Barium peroxide, for example, though it reacts readily with H_2SO_4 (SO_3) does not form a corresponding salt therewith. Hence it ought not to rank as a basic oxide. But barium peroxide reacts readily with the acidic oxide SO_2, giving as sole product the salt $BaSO_4$; therefore, from this point of view, the peroxide is a basic oxide.

When manganese dioxide is heated with potash in the presence of air, a salt K_2MnO_4 is formed. This is said not to correspond to the dioxide MnO_2, but to the trioxide MnO_3, and its formation under these conditions is not regarded as any proof of acidic properties in MnO_2. This seems to imply that the MnO_2 is first oxidised to MnO_3 by the oxygen of the air, and this acidic oxide then reacts with the potash to form the corresponding salt potassium manganate. But we have no proof that the MnO_2 does not itself first combine with the potash to form a corresponding manganite, which as soon as it is formed oxidises to manganate. Or, since K_2O is known to be peroxidised when heated in the air, it may be that the exceedingly unstable salt K_2MnO_4 results from the interaction of the acidic oxide MnO_2 and the basic oxide K_2O_2. But enough has been said to indicate some of the difficulties attendant on the attempt to rigidly classify the oxides.

Before treating of salts, the subject of neutralisation

merits a passing notice. When an alkali in aqueous
solution is added to an aqueous solution· of a strong
acid in *just* the right quantity to destroy completely
both the characteristic properties of the acid and the
alkali—for the two sets of properties disappear simultaneously—the acid is said to be neutralised by the
alkali, and *vice versâ.*

We might employ the disappearance of any one of the
characteristic properties of acids or alkalis as an index
of the realisation of complete neutralisation, but the
property hitherto almost universally selected for this
purpose is the reddening action of acids, and the blueing
action of alkalis on the vegetable colour litmus—the
tinctorial matter of a certain species of lichen. The
normal hue of this substance is purple, but it turns
red when treated with acids, and blue when treated
with alkalis. Suppose then we have an acid solution
which we wish to exactly neutralise with a solution of
alkali, we add a few drops of litmus solution to the
acid, and then add the alkali solution cautiously till
the red solution containing the acid becomes purple.
The neutralisation is then considered to be exact,
whereas, if too much alkali had been added, the solution would be blue, if too little, red. Processes of this
kind are known generally as processes of titration, for
they are usually carried out with, or rather involve
somehow, solutions whose strength or *titre* is known
beforehand; and colouring matters like litmus used as
aids to exact titration are called indicators. But the
introduction of numerous coal-tar products which give

colour changes with acids and alkalis[1] leads us to ask if the neutral tint of litmus is an index of exact neutralisation, *i.e.*, the presence of acid and alkali in *precisely* the quantities theoretically necessary for complete combination.

These new artificially prepared indicators are so sensitive that they do not show intermediate neutral tints as is the case with litmus. They not only differ among themselves in sensitiveness to, but also in their qualitative attitudes towards, one and the same solution. Thus, potassium sulphite solution is neutral to phenolphthalein, but changes violet litmus to blue just as an alkali does. Copper sulphate solution is acid as tested by litmus, but neutral in its behaviour to lacmoid. Saliva, which is normally neutral to litmus, is strongly alkaline to lacmoid and acid to turmeric; and

---

[1] Among these new indicators may be mentioned—

Phenolphthalein : with acids, colourless ; with alkalis, purple red.
Methyl orange : with acids, pink ; with alkalis, pale yellow.
Lacmoid : with acids, red ; with alkalis, blue.
Rosolic acid : with acids, pale yellow ; with alkalis, violet red.

The behaviour of a substance in the *rôle* of indicator, seems to be determined by the balance of acidic and basic characters in its molecule. If the acidic properties are relatively strong, the indicator is especially sensitive to bases, and can be successfully used in the titration of salts formed from weak acids, such as carbonates, sulphides, borates, &c. For carbonic, sulphydric, and boric acids, which are liberated during the process of such titrations, are unable to affect the colour of the indicator. If, on the other hand, basic characters have the predominance in the molecule, then the indicator, being very sensitive to acids, is useless for titrations wherein even such feeble acids as carbonic, sulphydric, &c., are produced. The differences in sensitiveness of these indicators have been applied to the simplification and shortening of many processes of volumetric analysis.

so on. Indicators, then, differ among themselves in their testifications as to acidity, alkalinity, and neutrality, and there is no reason why the indications of litmus should be accepted in preference to those of other colouring matters.

Salts.—The term salt did not always have its present connotation. It was at one time loosely applied to all substances which tasted like sea-salt, were easily soluble in water and recoverable from their solutions by evaporation. Of these characteristics solubility was regarded as the most important; hence among the alchemists we find an acid referred to as *sal acidum*, an alkali as *sal alkali*, while a salt proper was distinguished as *sal salsum*. It was Rouelle who, totally disregarding solubility relations, appropriated the term salt for the product of the interaction of an acid with a base or metal. The essential feature of an interaction of this kind is the replacement, total or partial, of the replaceable hydrogen of the acid by equivalent quantities of metals. If the whole of the replaceable hydrogen of an acid is displaced by a metal, the resulting salt is in general called a normal salt. Thus Na_2CO_3 (from the acid H_2CO_3Aq) $CuSO_4$ (from H_2SO_4) and Na_3PO_4 (from H_3PO_4) are all normal salts. The normal salts formed by the interaction of the strong acids (HCl, H_2SO_4, HNO_3, &c.) and the strong bases (KOH, NaOH) were the first to be investigated. As such salts (*e.g.*, K_2SO_4, NaCl, KNO_3, &c.) were all neutral to litmus, it was once customary to regard as synonymous the terms neutral salt and

normal salt.   But it is now known that many normal
salts are not neutral.   When the metal of a weak base,
e.g., $Fe(OH)_3$ replaces all the hydrogen of a strong acid,
e.g., HCl, a normal salt with acidic properties results,
e.g., $FeCl_3$.   On the other hand the interaction of a
strong base, e.g., KOH and a weak acid, e.g., H_2CO_3
gives a normal salt with alkaline properties, e.g.,
K_2CO_3.

If only a portion of the replaceable hydrogen of the
acid is replaced by metal, as in the compounds $NaHSO_4$,
$NaHCO_3$, Na_2HPO_4, the resulting salts are called acid
salts.   This title does not necessarily imply the possession of any other acidic property than the presence of
one or more hydrogen atoms replaceable by metals; as
we shall see presently, it does not always imply even
this characteristic.   It denotes a particular composition
rather than a definite set of properties.   Thus the
slightly alkaline body Na_2HPO_4 is in terms of our
definition an acid phosphate of sodium, while acid
sodium carbonate $NaHCO_3$ possesses very pronounced
alkaline properties.[1]

But it is necessary to enlarge our ideas of acid salts.
If acid salts were nothing else than acids in which
part of the replaceable hydrogen is substituted by an
equivalent quantity of metal, then it follows that a
dibasic acid like carbonic acid (H_2CO_3Aq) should be
capable of forming only one acid sodium salt $NaHCO_3$.

---

[1] Acid salts containing replaceable hydrogen and formed from dibasic
acids are often distinguished from the normal salts formed from the
same acid and base by the prefix bi-.   Thus $NaHSO_4$ is bisulphate of
sodium; $NaHCO_3$ bicarbonate of sodium, and so on.

But there is another way of looking at, and defining, acid salts. In the normal sodium salt of carbonic acid Na_2CO_3 or $Na_2O.CO_2$

the acid oxide : the basic oxide : : one reacting weight : one reacting weight.

In the acid salt $2NaHCO_3$ or $Na_2O.2CO_2.H_2O$

the acid oxide : the basic oxide : : two reacting weights : one reacting weight.

In other words an acid carbonate of sodium is a salt in which the ratio $\dfrac{\text{reacting weights of acidic oxide}}{\text{reacting weights of basic oxide}}$ exceeds the value unity. This is a wider definition than the one previously given, in that it does not point to any limit for the number of possible acid salts. It prepares us for the statement that a second acid carbonate of sodium exists. This is the so-called sesquicarbonate of sodium which has the composition $2Na_2O.3CO_2.3H_2O$.

Every oxy-salt can be conventionally regarded as a compound of basic with acidic oxide. Let us suppose that in the normal salt of a given metal the ratio

$$\frac{\text{reacting weights of acidic oxide}}{\text{reacting weights of basic oxide}} = n$$

then if other salts formed from the same acid and base or metal exist in which the above ratio has a greater value than n, they must be regarded as acid salts.

We see then that it is not necessary that every acid salt should contain replaceable hydrogen. Thus, in addition to the normal sulphate of antimony $Sb_2(SO_4)_3$ [$Sb_2O_3.3SO_3$], a sulphate having the composition $Sb_2S_4O_{15}$ [$Sb_2O_3.4SO_3$] is known. It is therefore in terms of our latest definition an acid salt, and as such it is commonly regarded, though it does not contain replaceable hydrogen.

Why then, it may be asked, should it not be called a normal salt, seeing that all the hydrogen of the acid from which it has been formed (H_2SO_4) has been replaced? The answer is, that by tacit consent only those salts are called normal which, in addition to the absence of replaceable hydrogen, can be represented as containing whole multiples of the radicles of the acid forming them.[1] Thus all normal sulphates can be represented as containing electro-positive element united with $n(SO_4)$; all normal carbonates will contain $n(CO_3)$; all normal phosphates $n(PO_4)$; and so on.

If the ratio $\dfrac{\text{reacting weights of acidic oxide}}{\text{reacting weights of basic oxide}}$ for a particular salt fall below the value n peculiar to the normal salt, then the salt is called a basic salt. In normal mercuric sulphate $HgSO_4$ [$HgO.SO_3$] $n = 1$; in the sulphate Hg_3SO_6 [$3HgO.SO_3$] n equals only $\frac{1}{3}$. This latter compound is therefore a basic sulphate of mercury. These basic salts constitute quite a large

---

[1] The molecular formula of acids may be conceived as made up of two parts; one part consisting of the whole of the replaceable hydrogen, the residue constituting the other part being called the acid radicle.

and important class of bodies. They are for the most part insoluble substances formed when excess of water acts on the normal salts formed from weak bases, or when excess of weak base is allowed to interact with acid and a strong acid.

From the fact that there are several acids which do not contain oxygen and have therefore no corresponding acidic oxides, it follows that our definitions of acid and basic salts in terms of the ratio

$$\frac{\text{reacting weights of acidic oxide}}{\text{reacting weights of basic oxide}}$$

lacks generality. We can, however, regard *all* acid and basic salts from one common standpoint, provided we apply a vaguely extended meaning to the word neutralisation. When a weak base interacts with a strong acid, the resulting normal salt as we have seen has in general distinctly acid properties. In virtue of these properties it can interact with more basic oxide, producing what we may conceive of as a nearer approach to absolute neutralisation in the form of a basic salt. According to this view, the weaker bases should show themselves particularly prone to form basic salts with strong acids; and this is known to be actually the case.

Again, when a strong base interacts with a weak acid, basic properties predominate in the normal salt, in virtue of which it is capable of further interaction with more acid or acidic oxide, with the realisation of more perfect neutralisation in the form of an acid salt.

Thus it follows that acid salts are formed chiefly from the strong bases.

It has been customary to regard the excess of acid in an acid salt over and above that necessary for the production of a normal salt, as less intimately combined or associated with the base than that portion of the acid which just suffices for conditions of normality. Similar remarks apply to excess of base in basic salts. In conformity with these unsubstantiated views, acid and basic salts are often represented as *molecular compounds* of normal salt and acid on the one hand, and of normal salt and base on the other. Thus basic bismuth chloride is sometimes written $Bi_2O_3.BiCl_3$ [3 $BiOCl$]; sesquicarbonate of sodium, according to this method of representation, becomes $2Na_2CO_3.H_2CO_3.2H_2O$, and acid sulphate of potassium, $K_2SO_4.H_2SO_4$.[1]

But if normal salts still retain acid and basic functions, enabling them now to combine with more base, now with more acid, the question naturally arises, why should not normal salts with residual acidic functions combine with normal salts possessed of residual basic functions? This question introduces us to the class

---

[1] According to this view of the matter, the so-called hyperacid salts formed by monobasic acids, *e.g.*, $KCl.HCl$ or $KHCl_2$, $KF.HF$ or KHF_2 &c., ought not to be distinguished from the ordinary acid salt such as $Na_2SO_4.H_2SO_4$. The term "hyperacid salts" for these compounds is an outgrowth of the primitive views on acid salts, which represented them as acids in which only part of the replaceable hydrogen is replaced by metals. According to these views it is, of course, impossible for a monobasic acid to form acid salts; and hence the introduction of the term "hyperacid salts" to meet the necessities of such cases as $KHCl_2$, KHF_2 &c.

of double salts which is at present receiving a good deal of attention.

Normal sulphate of sodium, Na_2SO_4, exhibits residual basic properties in its interaction with sulphuric acid to produce the acid sulphate of sodium $Na_2SO_4.H_2SO_4$ (2 $NaHSO_4$). To zinc sulphate, by reason of the existence of such basic salts as $SO_3.2$ ZnO, $SO_3.4$ $ZnO,.2$ H_2O, $SO_3.6$ $ZnO.10H_2O$, and $SO_3.8$ $ZnO.2$ H_2O must be ascribed residual acidic properties. What wonder then that basic Na_2SO_4 combines with acidic $ZnSO_4$, forming the double sulphate $Na_2SO_4.ZnSO_4.4H_2O$?

The double halides are an especially important class of bodies, and much is to be hoped from their further investigation. Their formation and existence admit of a provisional explanation similar to that just employed in connection with the double sulphates.

By reason of the existence of the unstable hyperacid salts having the general formula MX.HX, where X is a halogen and M the metal of an alkali, slight basic properties must be allowed to the alkaline halides. But the normal salts formed by the interaction with the haloid acids of the weaker bases (*i.e.*, the oxides and hydroxides of the heavy metals) have a surplus of acidic characters. Hence we have a large series of compounds, called double halides, of the general formula n (MX) m (M'X$_x$) where MX represents alkaline halide and M'X$_x$ the halide of some element other than the metal of an alkali. The following are a few examples, picked at random, of such double halides $BeCl_2.2KCl$,

K

$MgF_2.NaF$, $PbI_2.3NH_4Cl$. In accordance with the manner of regarding double salts here developed, it is found that these are most readily formed by halides corresponding to those oxides which in turn most readily form basic salts. Attempts have been made from time to time to remove the so-called double halides from the ill-defined and artificial class of *molecular compounds* by assigning to them normal unitary *atomic* structures.[1] These attempts necessitated certain extensions of our ideas of valency (see next chapter) which have not met with general acceptance; and until some more satisfactory theory of the structure of the double halides is framed, these bodies will, in all probability, continue to be empirically regarded as molecular compounds of the simple metallic halides from which they are prepared.

Inorganic compounds are not all included under the titles acids, bases, and salts. There is in addition a large and rapidly growing number of little-investigated binary bodies, such as the phosphides, the borides, the nitrides, the silicides, the carbides, the selenides, the tellurides, the arsenides, and the antimonides. Most of these substances (which may for the present be classed . together under the heading *indifferent bodies*[2]) are formed by direct union of the elements at high tem-

---

[1] For the distinction between molecular and atomic compounds see p. 155.

[2] The oxide NO together with several hydrides, *c.g.*, SiH_4, $P_2'H_4$, would also fall under this category of indifferent bodies. For an exhaustive list of these indifferent bodies, the reader is referred to Ramsay, *System of Inorganic Chemistry*, p. 497 *et seq.*

peratures and are decomposed by water. Though in several cases the phosphides, &c., of an element M are of the same type as the phosphides, &c., of the element hydrogen (*e.g.*, P_2H_6 and $P_2Zn''_3$), yet they cannot be regarded as salts of the latter. For the hydrogen compounds do not exhibit the most universal of acid properties—the possession of hydrogen atoms directly replaceable by equivalent quantities of other elements.

CHAPTER VI.

ISOMERISM AND MOLECULAR ARCHITECTURE.

WITH respect to the arrangement of the simple atoms in the compound atom (molecule) the earlier chemists did not concern themselves. The determination of the mere composition and formula of the compound atom was a problem more than sufficient for their day. But during the years 1820–25 the discovery and investigation of cyanic, fulminic, and cyanuric acids forced chemists to face the question of the mutual relations of the atoms in the molecule. These three acids were found to have identically the same percentage composition, and yet the properties of the three bodies differed most pronouncedly.

So foreign did the possibility of the existence of two or more different bodies of the same percentage composition appear to the early chemical philosophy, that Berzelius for long refused to admit it. Finally, however, the accumulation of well-authenticated instances of the phenomenon, among which the different tartaric acids may especially be mentioned, not only forced Berzelius to an admission, but led him to introduce the general term "isomerism"[1] for the class of facts under observation.

[1] Much confusion exists in regard to the precise use of the terms isomerism, metamerism, &c. Some, regarding identity of percentage

What explanation of isomerism could there possibly be other than a difference in atomic arrangements in the molecules of isomeric bodies? Henceforth it was acknowledged that the properties of a substance depend not only on its composition but also on the atomic architecture of its molecules. Now the architecture of two molecules having identical atomic compositions can differ in two ways. The relative positions of the atoms remaining the same, their distances apart may differ; or the atomic distances being constant or without influence, the relative positions may differ. It is believed, not without strong evidence, that differences in the relative positions of the atoms constitute the sole determining cause of isomerism; differences in atomic distances, if indeed such exist, being without influence.

composition as the sole condition of isomerism, divide isomers into the two classes polymers and metamers; polymers having different molecular weights, metamers the same molecular weight.

Others again give two meanings—a wide and a restricted one—to the term isomer. As before, all bodies having the same percentage composition are isomers in the wide sense. Such isomers are then classified into (a) bodies of different molecular weights, (β) bodies of the same molecular weight. Bodies of the former class are polymers. The latter class is further subdivided into (a') bodies of the same type or isomers in the restricted sense, and (β') bodies of different types or metamers.

Throughout this chapter the terms will be used in accordance with Berzelius' first suggestion, viz. :—

Polymers—bodies of the same percentage composition and different molecular weights.

Isomers—bodies of the same percentage composition and the same molecular weight.

Metamers—closely related isomers which are capable of very readily changing one into another.

(See article "Isomerism" in *Watts' Dictionary of Chemistry*.)

At any rate, all known cases of isomerism can be explained in terms of differences in atomic arrangement alone.[1]

Two, and only two isomers of the formula C_2H_6O are known; all attempts to produce a greater number have been futile. But it is obviously possible to arrange the nine atoms in a multitude of different ways. Why then does not a multitude of isomers exist? There must be a something limiting the possible groupings of a number of atoms in a molecule, whose cause must surely be sought for in the nature of the atoms themselves.

To take an analogy. Imagine a number of athletes of widely varying strength or holding power. So long as we pay no regard to their holding powers it is evident that we can arrange these athletes in a large number of different groups—in each group the relative positions of the athletes being different.

But the number of possible arrangements is very much diminished as soon as we make the stipulation that the groups are to be conditioned by, and exist in virtue of, the holding powers of the individual athletes —groups such as the "human trees" which so frequently form one of the items in a circus programme.

So it is with the atoms in a molecule; they are not thrown together "higgledy-piggledy" and without any dependence on the intrinsic peculiarities of the atoms,

---

[1] The distances between those atoms of a molecule which have "strong affinities" for each other may be a determining cause in isomeric change. But *the isomeric change from one body A to another body B* is a very different thing from *the isomerism of A and B*. (See p. 183).

but the molecule owes its continued existence to the definite holding or linking power of its constituent atoms. To this property or power of atoms the general term valence or valency has been applied.

As at the birth-time of the recognition of valency, no molecules were known containing per one atom of hydrogen more than one atom of any other element X (*i.e.*, as there were no bodies of the type HX_n), and as the hydrogen atom had already been chosen as the standard for atomic weights, it seemed well to make it also the standard for valencies. Any atom X which held in combination one atom of hydrogen, and formed a molecule HX, was called a monovalent atom. The existence of the molecule H_2Y established the divalency of Y, and so on.

It will readily be seen that valency is nothing else than a special name given to equivalency when this is applied to atoms, and it is not therefore surprising that the same difficulties as beset the attempt to establish an equivalent system appear again in the later attempts to fix the valencies of the elementary atoms.

Only a few elements combine with hydrogen to form gasifiable molecules, and had we to rely only on hydrides, our knowledge of the valencies of the atoms of the elements would be very incomplete. Fluorine, chlorine, bromine, and iodine, however, combine atom for atom with hydrogen, hence they are univalent atoms, and may be used as middlemen to fix the valencies of elements which do not form hydrides. Unfortunately the valency of an atom as derived from a study of the halides of the element does not always agree with the valency as fixed by the hydride, *e.g.*, PH_3, PF_5. Moreover, many elements form at least two gasifiable halides with the same halogen, *e.g.*, $HgCl$, $HgCl_2$. Which particular halide is to determine the valency of the atom in question? Can we, indeed, correctly speak of *the* valency of an atom?[1]

This question has given rise to two opposing schools. One school, vaguely referring the phenomena of valency back to some objective attribute of the atom, asserts the necessary constancy of valency. The valency of an atom is, according to this school, as constant as its

---

[1] From a study of the periodic law, Mendeléeff arrived at the following generalisation: The sum of the "equivalents" of oxygen and hydrogen with which a single non-metallic atom is combined in its highest salt forming oxide on the one hand, and in its maximum hydride on the other, is constant for all non-metallic atoms, and is equal to 8. Thus, the highest salt-forming oxide of carbon is CO_2; the maximum hydride of carbon is CH_4. In each case the carbon atom is combined with four "equivalents," and $4 + 4 = 8$. The pairs of compounds PH_3, P_2O_5; AsH_3, As_2O_5; TeO_3, TeH_2, illustrate the same principle. It is obvious that by "equivalents," Mendeléeff here means something very much akin to valencies; and in this connection he asks why the valency of an atom should be gauged by its hydrides rather than by its oxides. It should be noted that Mendeléeff's generalisation is founded on the assumption that peroxides (*i.e.*, oxides of higher types than the so-called group oxides) are not strictly salt-forming—do not form corresponding salts. The recent isolation of salts $M_2S_2O_8$ (where M = monovalent atom) corresponding to sulphur peroxide S_2O_7 is at variance with this assumption. The sum of the "equivalents" of sulphur in S_2O_7, and SH_2 is 9, and not 8. Some other criterion than the power of forming corresponding salts must lie at the basis of a classification of oxides adapted to Mendeléeff's generalisation. Further, the partiality of the generalisation, including as it virtually does only the non-metals, is against it. See Mendeléeff's *Principles of Chemistry*, vol. ii., appendix 1.

weight, although under certain conditions it is not wholly in evidence. Among modern chemists, van't Hoff may be cited as an adherent of this school. He assumes that the attractions in virtue of which the atoms of a molecule hold together, are of the same order as the gravitative attractions of ponderable masses. This being the case, he shows that the intensity of the attraction over the surface of an atom would be constant only if the atom were truly spherical. If, on the contrary, the atom had any other figure than the sphere, then at certain points on its surface the attractive forces would have maximal values—these values being unequal *inter se*. According to van't Hoff's view, then, the valency of an atom expresses the total number of points of maximal attraction—this number being dependent on the form of the atom. As the form of the atom is presumably constant, the valency is also constant. If, however, under certain conditions the movements of the atoms conditioning the temperature of the gas become so energetic that only the higher maxima of attraction are powerful enough to come into effective operation, the atom will apparently possess a valency lower than its actual valency.[1]

[1] In view of the fact that Faraday's law of electrolysis can be stated in terms of valency data, it may seem somewhat surprising that no satisfactory electrical theory of the cause of valency has appeared. Yet it should be remembered that valency had its origin, and finds its chief application, in organic chemistry, and the great majority of organic compounds, *i.e.*, all organic compounds which are not salts nor acids, are non-electrolytic. In electrolysis the positive electricity may be regarded as conveyed through the electrolyte from anode to kathode by the metallic atoms, the negative electricity from kathode to anode

The opposing school, most non-committal and un-imaginative, does not attempt to theorise on, or explain in any way, the phenomena of valency. It does not speak of the absolute valency of any atom, but only of the maximum known valency of an atom or, less general still, of the valency of an atom in a particular compound.[1]

The history of the school of constant valency is an instructive one. Its attempts to vindicate its tenets gave rise not only to such familiar distinctions as those involved in the terms *saturated and unsaturated bodies, atomic and molecular compounds*, but also to the celebrated so-called theory of bonds.

by the non-metallic atoms or acid radicles. Atoms and radicles in their capacity as carriers of electricity during electrolysis are called ions. Faraday's law can be stated in the following form :—The electrical carrying power of an ion is directly proportional to its valency. In other words, a quantity of electricity, positive or negative, which is conveyed through a salt solution by an *n*-valent atom or radicle, would require for its convection *n* monovalent atoms or radicles. If a current of electricity be passed through two electrolytic cells in succession, the amount of electricity passing through each cell in a unit of time is the same. If in the first cell we have a salt of a divalent metallic atom, in the second a salt of a monovalent metallic atom, then for every atom (or what is the same thing, for every atomic weight expressed in any mass unit) of the divalent metal deposited in the first cell we shall have two atoms (two atomic weights in the same mass unit) of the monovalent metal deposited in the second cell. See Lodge, *Modern Views of Electricity*, p. 72 *et seq.*

The table given in note 1, p. 159, could quite well be constructed from quantitative electrolytic determinations alone.

[1] Apropos of the attitudes of these rival schools, Ostwald writes :— " Fragt man nach dem Wege, auf welchem eine Entscheidung zu treffen wöre, so kann eine solche nur auf Grundlage einer bestimmten, wohl begründeten Hypothese über die Natur dessen, was wir Valenz nennen, erlaugt werden."

The nitrogen atom combines with three atoms of hydrogen to form NH_3; therefore nitrogen is trivalent. In order to express this fact clearly and succinctly, the notation $-\text{(N)}-$ was employed. At first the lines proceeding from the circle enclosing the symbol for nitrogen were regarded in their true light—simply as a species of chemical shorthand; but soon they came to be regarded as symbolising some objective characteristics of the atoms called bonds or units of affinity. The nitrogen atom was trivalent, *because* it was possessed of three bonds or units of affinity. It did not hamper the advance of the school, that for long no satisfactory answer could be given to the question, what is the nature of a bond or affinity unit? But chemists are not the only people who have at times deceived themselves into the belief that to name the unknown is to explain and progress.[1]

But if nitrogen is trivalent, and hydrogen and iodine both monovalent, how, it was asked by the opponents of the school, could the existence of the molecule NH_4I be explained? In terms of the difference between atomic and molecular compounds, was the answer. Only in the former class of bodies, it was said, is valency operative. Ammonium iodide is a loose molecular compound of the two atomic compounds NH_3 and HI, in the former of which the nitrogen atom displays its customary

---

[1] Thanks to van't Hoff's theory of valency, we may now, without being accused of a mere glossing of our ignorance, employ the term "bond." Some such term, properly understood, is not only generally useful, but it is absolutely indispensable in describing the results of the new stereo-chemistry (*vide infra*). The word bond is much to be preferred to its once synonymous term "unit of affinity."

trivalency.   The attraction (or energy degradation), in virtue of which the atomic compounds HI and NH_3 mutually hold each other in combination, is not only independent of, but of a quite different order from, the attractions associated with the valencies of the atoms, and instrumental in holding together the parts of atomic compounds.   The former attractions are attributes of the molecules as wholes, and are of a physical rather than a chemical nature.

However, experiments on the substituted ammonium iodides (NH_4I in which the H atoms are replaced by different monovalent organic radicles) showed in this special case the artificiality of the distinction between atomic and molecular compounds, and the necessity of regarding ammonium iodide as a true atomic compound in which the nitrogen atom, holding in direct combination the four atoms of hydrogen and the one atom of iodine, must be pentavalent.[1]

The school of constant valency was bound to recognise the justness of these conclusions, to which it immediately adapted itself by instituting a distinction between saturated and unsaturated compounds, maintaining the while that the nitrogen atom was constantly pentavalent instead of trivalent, as hitherto upheld.

Here will be observed a slight departure from the original signification of valency ; the valency of an atom is now measured by the number of bonds it possesses.

---

[1] The behaviours of such presumed inorganic molecular compounds as $KTlI_4$ [$KI.TlI_3$], $K_3SbBr_3Cl_3$ [$SbBr_3.3KCl$], &c., are also not in harmony with the deductions which necessarily flow from the admission of molecular combination as distinguished from true atom linking.

It at first sight seems more definite than the original signification in that a cause for the phenomena is assigned. But phenomena which are not understood are not cleared up by the mere assigning of causes whose natures are themselves darkly obscure. It is worthy of remark that till van't Hoff's time no satisfactory explanation was advanced of the reason why an atom forms sometimes a saturated, sometimes an unsaturated molecule.

Conceptions of valency had their origin in the study of carbon compounds, and it is in the field of carbon compounds that the developed theory to-day finds its most successful applications.[1] The reasons for this are not far to seek. The carbon atom is peculiar in that its valency determined from *all* its highest forms of combination containing only a single carbon atom in the molecule, is constant. The molecules CH_4, $CHCl_3$, $CHBr_4$, CCl_4, CS_2, and CO_2 all point to the tetravalency of carbon.[2] Again, carbon compounds are for the most part easily vaporised, and hence the molecular

[1] It should be particularly noted that organic chemistry, though largely indebted to, is nevertheless quite independent of, the theory of valency. Mendeléeff has shown that the application of Newton's third law of motion in the form in which it appears when regarded as a corollary of the first law, combined with the principle of substitution, is capable of effecting all that the doctrine of valency really effects, viz., the limitation of the number of isomers possible for a given atomic complex, and the provision for each isomer of an appropriate molecular ground plan. (See Mendeléeff, *Nature,* vol. xl. No. 1032, and Carnegie, *American Chemical Journal,* vol. xv. No. 1.)

[2] It may be that some day this series will be rendered complete by the discovery of a substance having the molecular composition CN_2; the group N_2 being tetravalent, as in the molecule of hydrazine $H_2 = N_2 = H_2$.

weights of the majority of carbon compounds are known. On the other hand, of only some sixty inorganic compounds of the type MX_n are the true molecular weights known, and it is impossible to draw unequivocal conclusions regarding the valency of an element M which does not from gasifiable compounds of the type MX_n, where X represents some monovalent atom or group. No compound of sodium has been vaporised. Sodium chloride *may* have the molecular composition NaCl, in which case sodium is a monovalent atom; but there is no *à priori* reason why the molecular composition of salt should not be $Na_2Cl_2 \ldots Na_nCl_n$, in which cases the sodium atom may be respectively di- tri- $\ldots n \ldots (2n - 1)$- valent.

Further, of the elements with gasifiable compounds of the type MX_n none except silicon, which is very closely related to carbon, shows a constant valency as X is varied. Thus, the phosphorus atom is trivalent in phosphine PH_3, and pentavalent in phosphorus pentafluoride PF_5. Nor, by reason of the comparative simplicities of the molecules of inorganic bodies, the thorough and so to speak annihilatory nature of the changes they undergo, and the absence of marked cases of isomerism among them,[1] are questions rela-

---

[1] The only well-marked cases of isomerism (it *may*, however, be polymerism) in inorganic chemistry are afforded by the so-called inorganic amines—the numerous and complex bodies which are formed by the action of ammonia, under varied conditions, on the salts of such metals as platinum, cobalt, chromium, and rhodium. See Ramsay, *System of Inorganic Chemistry*, p. 524, *et seq.* The structural formulæ which have in some cases been assigned to these inorganic isomers are open to great doubt and uncertainty.

ting to structure of so much importance in inorganic chemistry as they are in the domain of carbon compounds.

For all these reasons the theory of valency admits of much more definite and productive application to carbon compounds than to inorganic bodies; indeed it may seriously be questioned whether the necessarily loose and gratuitous applications of considerations of valency to inorganic compounds have not impeded the progress of the science at large.[1] That such applications brought discredit on the theory itself will be admitted by most.

[1] The valency data of greatest practical importance in inorganic chemistry are here tabulated—

I. *Metals and basic radicles*—

    (α) *monovalent*, K, Na, Li, Rb, Cs, NH_4, Ag, Hg^{ous}, Cu^{ous}, Au^{ous}, Tl^{ous}, BiO.

    (β) *divalent*, Ca, Sr, Ba, Pb, Mg, Zn, Cd, Hg^{ic}, Cu^{ic}, Sn^{ous}, Fe^{ous}, Cr^{ous}, Mn^{ous}, Co, Ni, Be, Pt^{ous}, UO_2.

    (γ) *trivalent*, Au^{ic}, Al, Bi, Fe^{ic}, Cr^{ic}, Tl^{ic}.

    (δ) *tetravalent*, Sn^{ic}, U^{ous}, Pt^{ic}.

II. *Non-metals and acidic radicles*—

    (α) *monovalent* – Cl, – Br, – I, – F, nitrites – NO_2, nitrates – NO_3, hypochlorites – ClO, chlorates – ClO_3, cyanides – CN, cyanates – CNO, thiocyanates – CNS, metaphosphates – PO_3, perchlorates – ClO_4.

    (β) *divalent* – O, – S, sulphites – SO_3, sulphates – SO_4, thiosulphates – S_2O_3, carbonates – CO_3, silicofluorides – SiF_6, chromates – CrO_4, dichromates – Cr_2O_7, orthomolybdates – MoO_4, tungstates – WO_4, metaborates – B_2O_4, pyroborates – B_4O_7, metasilicates – SiO_3.

    (γ) *trivalent*, phosphites – PO_3, phosphates – PO_4, ferricyanides – $Fe(CN)_6$, arsenites – AsO_3, arseniates – AsO_4, orthoborates – BO_3.

    (δ) *tetravalent*, ferrocyanides – $Fe(CN)_6$, pyrophosphates – P_2O_7, orthosilicates – SiO_4.

By means of this table, the formula of any common inorganic salt can be found. Place the symbols for the metal and acid radicle of the

The chief services that the theory of valency has rendered the chemistry of carbon compounds have been in the domain of isomerism. The theory enables us to predict with absolute certainty the number of isomers consistent with a given molecular composition. The general idea one is apt to glean from much of the literature on the subject is that valency lies at the basis of the so-called constitutional or structural formulæ. This is not the case. What does lie at the basis of structural formulæ is merely the conception of definite atomic linkage—a conception of lower order than that of valency. It is quite possible to symbolise succinctly the chemical relations of a body—to give it a structural

salt in juxtaposition; then multiply one or other of these symbols, or it may be both, so that the metallic atoms and acid radicles become equivalent. Express these multiples in small subscripts placed after the respective symbols, and the result is the formula required. Thus calcium phosphate is $Ca_3(PO_4)_2$; for we must take three divalent atoms and two trivalent radicles in order to get a balance of the valencies of the two constituent parts of the salt.

The table is necessarily incomplete. Elements which form halides of several types, but no stable oxysalts, e.g., W and Mo, have been omitted. As regards the formulæ of oxides, the table is only available for those markedly basic oxides which form corresponding salts; it correctly gives the formulæ for such oxides as ferrous oxide, baric oxide, &c., but it does not help us in the cases of the magnetic oxide of iron, baric peroxide, &c. Further, it only gives the formulæ for normal salts; the complex poly-salts formed by weak acids such as boric, molybdic, tungstic, silicic, &c., are most simply represented, not as made up of metal and acid radicle, but of acidic and basic oxide, thus—n BO m AO, where BO is the oxide of a metal, AO the oxide of a non-metal, and m n are whole numbers. However, these complex salts can also be represented as derived from the normal salts. Thus the mineral serpentine usually written $2SiO_2.3MgO$, may also be regarded as a dehydrated acid magnesium ortho-silicate, $Mg_3H_2(SiO_4)_2 - H_2O = Mg_3Si_2O_7$. Obviously, the table does not inform us as to the possible number or the limit forms of such complex salts.

formula recalling its mode of formation, suggesting its decompositions, &c.,—by means of certain conventions respecting the written arrangements and groupings of the constituent atoms, and without in any way having recourse to the teachings of valency. Thus, the symbol

$$HO.\overset{O}{\underset{O}{S}}.OH$$

conveys much information concerning sulphuric acid, and yet is framed without any regard to the doctrine of valency. This symbol pictures the fact of the symmetry of the sulphuric acid molecule, and therefore the impossibility of the existence of isomeric forms of the derivatives $HRSO_4$ and $RR'SO_4$.

Further, knowing the equal value, substitutionally, of Cl and (OH), of H and K, &c., and also the readiness with which hydroxyl groups linked to the same atom interact, the following transformations are at once suggested by the symbol given—

$$Cl.\overset{O}{\underset{O}{S}}.Cl \quad Cl.\overset{O}{\underset{O}{S}}.OH \quad Cl.\overset{O}{\underset{O}{S}}.OK \quad KO.\overset{O}{\underset{O}{S}}.OK \quad \overset{O}{\underset{O}{S}}.O$$

To argue whether the sulphur atom is tetra- or hexavalent in sulphuric acid is profitless. The formulæ

$$HO-\overset{\overset{O}{\|}}{\underset{\underset{O}{\|}}{S}}-OH \qquad \overset{HO}{\underset{HO}{>}}S\overset{O}{\underset{O}{<}}$$

do not convey any more information about the properties of sulphuric acid than does the simple representation $HO.\overset{O}{\underset{O}{S}}.OH.$

L

Before proceeding to illustrate the applications of considerations of valency to organic bodies, it is first necessary to speak of the valencies of compound radicles. A compound radicle is a group of atoms which, like a single elementary atom (simple radicle) appears as an undecomposed whole throughout a large series of compounds or transformations. Like an atom, too, a compound radicle has a definite valency. These compound radicles, or stable atom complexes, play so conspicuous and important a *rôle* in the field of carbon compounds that they were once regarded as the true atoms of organic chemistry, which itself was supposed to be adequately defined as the chemistry of compound radicles. As random examples of these compound radicles we may cite ethyl C_2H_5, ethyoxyl C_2H_5O, hydroxyl OH, carboxyl COOH, and amidogen NH_2 —all monovalent radicles, for each combines with a single atom of hydrogen to form a gaseous molecule; the divalent radicles ethenyl C_2H_4, carbonyl CO, and the trivalent radicles, glyceryl C_3H_5 and methenyl CH.

Now, if the carbon atom can hold in combination four hydrogen atoms, it will also hold in combination four monovalent compound radicles, and therefore such molecules as

$$OC_2H_5 \qquad H \qquad NH_2$$
$$| \qquad | \qquad |$$
$$C_2H_5O - C - OC_2H_5 \quad HOOC - C - COOH \quad H - C - COOH, \&c.$$
$$| \qquad | \qquad |$$
$$OC_2H_5 \qquad H \qquad H$$

ought to, and do, exist.

Hence we are bound to give an extended significance to the statement—the carbon atom is tetravalent. Not only does it imply that the carbon atom combines with four atoms of hydrogen to form a gaseous molecule, but further that in any molecule a carbon atom is capable of directly interacting with four, but no more than four, other atoms, the particular valencies of these atoms being a matter of indifference. Thus in the ethyl ether of ortho-carbonic acid, $C(OC_2H_5)_4$, the heavily printed carbon atom directly interacts with four divalent oxygen atoms, each of which again directly interacts with a monovalent ethyl group. In glycocine

$$CH_2.NH_2.COOH$$

the heavily printed carbon atom acts on two monovalent hydrogen atoms, one trivalent nitrogen atom, and one tetravalent carbon atom—four atoms in all.

Let us now proceed to apply these considerations to isomerism. A body is discovered having the molecular composition C_2H_6O. Can it have isomers, and if so, how many? The answer returned by the valency theory is this. If the constituent atoms can be arranged in n different ways so that each of the carbon atoms never interacts directly with more than four other atoms, and each of the oxygen atoms with more than two other atoms, then n isomers, all told, are possible. As a matter of fact, it is found on trial that the atoms C_2H_6O can be arranged in only two ways, subject to the restrictive conditions just stated—

$$
\begin{array}{ccc}
\text{H} & \text{H} \\
| & | \\
\text{H} - \text{C} - \text{O} - \text{C} - \text{H} \\
| & | \\
\text{H} & \text{H}
\end{array}
\qquad
\begin{array}{cc}
\text{H} & \text{H} \\
| & | \\
\text{H} - \text{C} - \text{C} - \text{O} - \text{H} \\
| & | \\
\text{H} & \text{H}
\end{array}
$$

In strict accordance with this result, all attempts to prepare more than two isomers of the composition C_2H_6O have failed. The two known substances having this formula are methyl ether and ordinary alcohol.

The problem next rises, which of these formulæ are we to ascribe to methyl ether; which to alcohol? This is solved by the knowledge of the fact that one atom of hydrogen in the alcohol molecule differs from the remaining five in that it alone can be replaced by alkali metals. Now, in the left-hand formula all the atoms of hydrogen are similarly related to the molecule, but in the right-hand formula five atoms of hydrogen directly interact with carbon atoms (*i.e.*, are directly linked to carbon atoms), while the sixth hydrogen atom acts only indirectly on a carbon atom through the oxygen atom. Therefore the right-hand formula belongs to the molecule of alcohol, and therefore by exclusion the left-hand one represents methyl ether. Further, the syntheses and all the known transformations of methyl ether, and ethyl alcohol, are in strict conformity with, and indeed are deducible from, their appropriate symbols thus arrived at.

Although in many cases more isomers ought to exist

according to the teachings of valency than are at present known,[1] yet in no case is a greater number of isomers known than is provided for, and anticipated by, the valency theory.

Till within quite recent years this statement could not have been made.

It would seem to follow from the simple case already considered, that isomerism ought always to be accompanied by differences in chemical properties, these differences finding expression in, and being due to, the differing molecular architectures of the isomers. Yet bodies are known which, having the same molecular weight and the same composition, undergo precisely the same chemical transformations, and are yet quite different in such physical properties as crystalline form, solubility, action on polarised light, &c. Such bodies, on account of the identity of their chemical transformations, must be assigned identical formulæ—identical molecular architectures ; and for this reason they were for long regarded as forming exceptions to, or at least falling outside the pale of the province of the "theory" of valency. Hence they were relegated to a kind of suspense class, bearing the title physical isomers.

We now know, thanks to the researches of van't Hoff and Le Bel, that these are not exceptional phenomena, but are fully provided for by the theory of valency, if we do not limit the possible different

---

[1] For instance, nine heptanes having the composition C_7H_{16}, ought to exist according to the teachings of valency ; as yet, however, only four bodies having this composition have been isolated.

arrangements of atoms in space by the different arrangements possible in a plane.

It is not irrational to assert that two isomers.of the formula CH_2Cl_2 should exist.   For the formulæ—

while both in accordance with the dictates of valency, obviously differ.   In the right-hand formula the two chlorine atoms are not related to the rest of the molecule in the same way as they are in the left-hand formula.   Why should not this difference entail a difference in properties—a possible isomerism of the molecule CH_2Cl_2?   It evidently does not, for all attempts to prepare two different bodies of the formula CH_2Cl_2 (and Henry has made many such) have failed. We are therefore compelled to still further enlarge our conceptions of valency by the conclusion that there is really no difference between the above formulæ—that the order of cyclical arrangement of the atoms round the central carbon atom is, in effect, immaterial.   Consequently, no isomeric molecules of the types CHR'_3, $CHR'R''_2$, $CHR'R''R'''$, $CR'_2R''_2$, CR'_3R'', $CR'R''R'''R'''$, should exist.   In partial accordance with this conclusion, no isomers of the first five types are known, but, contrary to the conclusion, all the simplest of the so-called physical isomers conform to the sixth type,

$CR'R''R'''R''''$—possess what is called an asymmetric carbon atom.

Van't Hoff and Le Bel showed the way out of this difficulty by reminding us that molecules are tri- and not di-dimensional entities.

We get a spatial conception of molecules in accordance with the limitations of valency, if we conceive the carbon atom placed at the centre of a tetrahedron, its four "valencies" or "bonds" being directed towards the four solid angles a, b, c, d, of the tetrahedron.

So long as the monovalent atoms or radicles on which the carbon atom directly interacts are not all different, then it is impossible to place them at the angles a, b, c, d, so as to occupy in different dispositions different

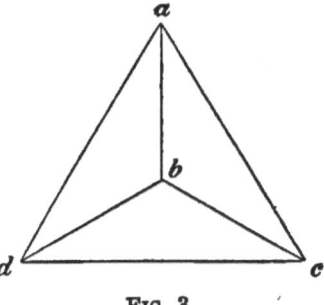

FIG. 3.

spatial relations to the central carbon atom. Let A stand for one of the dispositions, and B for another, and suppose the two tetrahedra interpenetrative, then it will always be possible, no matter the dispositions, so to place A inside B, or *vice versâ*, that the similar atoms or radicles attached to the central carbon atoms will fall together and occupy the same region of space. In short, the two dispositions will always be superposable.

But when the atoms or groups at the angles a, b, c, d, are all different from one another, then it is possible to get two dispositions which are non-superposable—the one disposition bearing to the other the same relation as does an object to its image in a plane mirror; the same relation as a left-hand glove bears to a right-hand one. In crystallography this would be called an enantiomorphous relationship. In other words, if in one molecule the atoms or radicles are regarded as disposed around the central carbon atom in a right-handed spiral, then in the other molecule the disposition is a left-handed spiral.

Let us illustrate this last case by the lactic acids which indeed first suggested to Wislecenus the necessity of introducing stereometric conceptions into the domain of molecular architecture.

Three lactic acids are known; they have all practically identical chemical properties, but differ from each other in their action on polarised light, and in the solubilities of their salts. Dextrolactic acid turns the plane of polarisation to the right; sarcolactic acid to the left; while fermentation lactic acid is optically inactive. The molecules of these lactic acids must, from their chemical relationships, have the structure

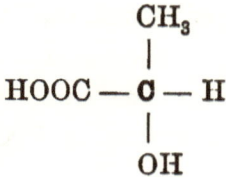

The heavily printed carbon atom in the above symbol is obviously asymmetric. Therefore there are possible

two different dispositions around it of the radicles it holds in combination, *viz.* :—

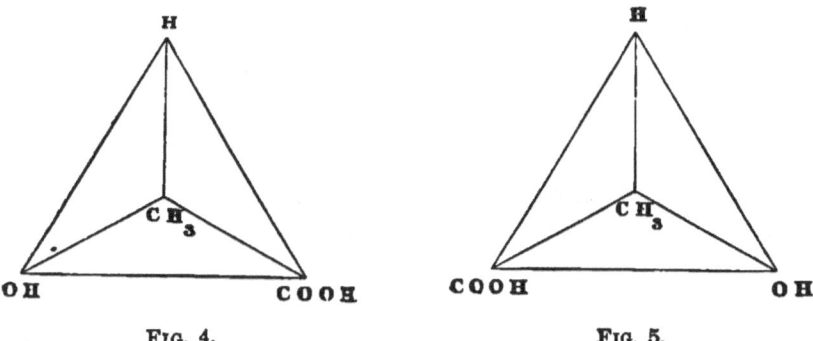

FIG. 4.   FIG. 5.

One of these dispositions must characterise dextrolactic acid; the other, sarcolactic acid. Yet in the present state of our knowledge it is impossible to definitely allocate them.

But what of the third lactic acid—the inactive modification? The theory at first sight does not seem to provide for it.

It has now been definitely proved that inactive or ordinary lactic acid is not a true chemical unit, but a mixture in compensating quantities of the dextro and laevo modifications. Its formula would be given by placing a plus sign between the above two figures.[1]

Attempts have recently been made to find some con-

[1] For the further development of stereometric ideas we would refer the reader to Marsh, *Chemistry in Space*, and Auwers, *Die Entwickelung der Stereochemie*.

Suffice it here to say that there are no known instances of physical isomersion of carbon compounds which do not receive full and adequate interpretation in terms of the attributes of the asymmetric carbon atom.

nection between the masses of the radicles, R^I, R^{II}, R^{III}, and R^{IV}, held in combination by an asymmetric carbon atom, and the nature and amount of the optical activity which, unless compensatory influences come into play, is invariably associated with asymmetric carbon atoms.

A regular tetrahedron has six planes of symmetry; a plane of symmetry bisecting each of the six interfacial angles of the figure.[1] Let us suppose that the centre of such a tetrahedron is occupied by a carbon atom, and that four atoms or radicles are disposed round the carbon atom so that their masses are concentrated at the apices of the four solid angles of the tetrahedron. So long as these four radicles are not all dissimilar, the centre of gravity of the whole system will be in one or other of the six planes of symmetry of the tetrahedron, and there will be no optical activity. But when all the radicles or atoms differ among themselves, the centre of gravity of the system no longer falls within any of the planes of symmetry, and optical activity makes its appearance. Let the perpendicular distances of the centre of gravity of the system in this latter case from the six planes of symmetry be d_1, d_2, d_3, d_4, d_5, and d_6. Then the product $P = d_1 \times d_2 \times d_3 \times d_4 \times d_5 \times d_6$ is, according to Guye, a measure of

---

[1] "A plane of symmetry may be defined as a plane which is capable of dividing a body into two halves which are related to each other in the same way that an object is to its reflection in a mirror. More exactly we may say : two objects or two halves of the same object are symmetrical with reference to a plane placed between them, when from any point of one object a normal to this plane, prolonged by its own length on the opposite side of the plane, will meet the corresponding point of the other object."—Williams, *Elements of Crystallography*.

the asymmetry of the molecule, and should therefore be proportional to the optical activity of the substance. If plus and minus signs be conventionally applied to the distances d_1, d_2, d_3, &c., according as they are measured from one side or the other of each plane of symmetry, P may be either positive or negative according to circumstances.

Guye has shown how P may be calculated beforehand for any meditated derivative of a given optically active substance. If the passage from an optically active body A to an optically active substitution derivative A′ would involve an increase in the value of P, then, according to Guye, we may predict that A′ will be more active optically than the parent substance; if the transformation would involve a diminution in the value of P, then the new body will be less active than the original one from which it was derived. Finally, if the passage from A to A′ is accompanied by a change of sign of P, then the parent substance and its derivatives will produce opposite rotations; one will be a dextro-rotatory substance, the other a lævo-rotatory substance.

In the following symbol for active amyl chloride, it can readily be shown that the centre of gravity of the system falls on the CH_2Cl side of the dotted plane of symmetry C_2H_5mH. If therefore the chlorine atom in the CH_2Cl group be replaced by a heavier atom, the

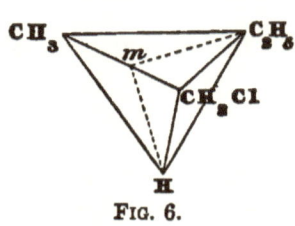

Fig. 6.

centre of gravity of the system must still remain on the same side of the plane C_2H_5mH, but will be further removed from it. Accordingly it is found that amyl bromide has higher optical activity than the chloride; and amyl iodide in turn higher activity than the bromide.[1]

Let us take another illustrative application of Guye's theory.

In the accompanying symbol for dextro-tartaric acid, the centre of gravity would

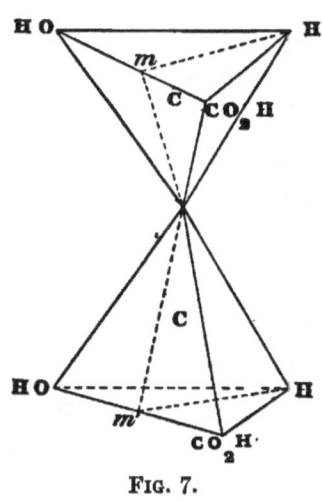

FIG. 7.

lie on the CO_2H side of the dotted plane Hmm'H. Let us now replace the H atoms of the hydroxyl groups by acetyl radicles (acetyl = CH_3CO). This substitution carries the centre of gravity over to the other side of the plane Hmm'H; and it is found as a matter of fact that diacetyl tartaric acid is lævo-rotatory.

Now let us form etherial salts of diacetyl tartaric

---

[1] The values are as follows :—

$$[a]_D = 1° \, 6' \text{ for the chloride.}$$
$$= 4° \, 24' \text{ for the bromide.}$$
$$= 8° \, 20' \text{ for the iodide.}$$

In these equations $[a]_D$ may be taken to represent the angle through which the plane of a polarised ray of sodium light would be turned if it were made to pass through a tube of $\frac{1}{10}$ square centimetre cross section, and of just sufficient length to hold exactly one gram of the substance under examination.

acid by replacing the H atoms of the carboxyl groups
with alkyl monovalent radicles R (*e.g.*, C_2H_5, C_3H_7, &c.).
This brings the centre of gravity of the system back
towards the CO_2R side of the plane H*mm'*H, and
consequently we find that the lævo-rotatory power
of the etherial salt diminishes as the weight of R
increases, until finally the rotation changes sign and
the higher etherial salts of diacetyl tartaric acid are
dextro-rotatory.

Diacetyltartaric acid	$[\alpha]_D$ =	$-$ 23·14°
Methyl-diacetyl-tartarate	,,	$-$ 14·29°
Ethyl ,, ,,	,,	$+$ 1·02°
Propyl ,, ,,	,,	$+$ 6·52°

Although many facts are in harmony with pre-
dictions founded on Guye's views of the relations
between optical activity and molecular asymmetry,

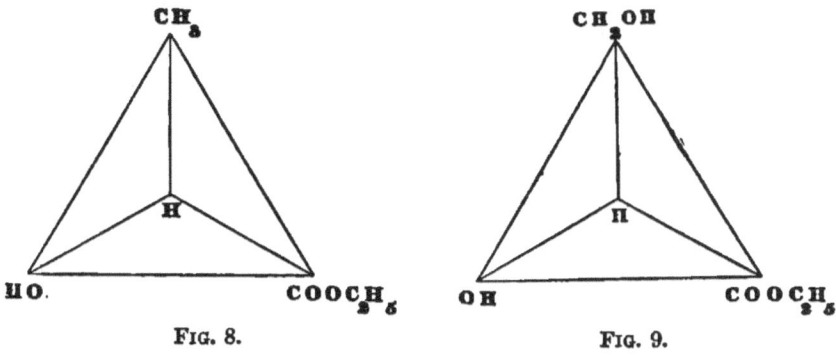

FIG. 8. FIG. 9.

yet isolated instances of exceptions to his generalisa-
tions are not wanting. Thus from the accompanying
diagrams it is easy to see that ethyl glycerate (Fig. 9)

should have higher optical activity than ethyl lactate
(Fig. 8); but as a matter of fact the optical constants
for these two bodies are as follows :—

$$\text{Ethyl lactate} \quad [\alpha]_D = -\ 14\cdot19°$$
$$\text{Ethyl glycerate} \ [\alpha]_D = -\ \ 9\cdot18°$$

Again, such a body as ethylic diacetyl glycerate

$$CH_2O\ .\ C_2H_3O$$
$$H\ .\ \overset{|}{C}\ .\ O\ .\ C_2H_3O$$
$$\overset{|}{C}\ .\ OOC_2H_5$$

should be optically inactive if mass were the only factor
determining optical asymmetry; for the two isomeric
radicles $CH_2\ .\ O\ .\ C_2H_3O$ and $COOC_2H_5$ have equal
masses, and when this is the case P must be zero. For
if a, b, c, d represent the masses of the radicles held in
combination by an asymmetrical carbon atom in such
a way that these masses are concentrated at the apices
of a regular tetrahedron, then one of the factors in
determining the value of P is—

$$\frac{(a\ -\ b)\ (a\ -\ c)\ (a\ -\ d)\ (b\ -\ c)\ (b\ -\ d)\ (c\ -\ d)}{(a\ +\ b\ +\ c\ +\ d)^6}$$

It is clear that if any two of the masses a, b, c, d
become equal in value, then P—the product of
asymmetry—must also become zero. When this is the
case, the optical activity which is supposed to accompany asymmetry in Guye's sense, should disappear.
Hence it would appear that in addition to mere mass

relationships, considerations of structure must also find a place in future attempts to quantitatively link together optical activity and molecular asymmetry.

So far we have only dealt with what were once regarded as exceptional isomerisms among saturated carbon compounds. We now turn to peculiar cases of isomerism presented by unsaturated carbon compounds of ethylenic and acetylenic types, *i.e.*, compounds in whose molecules one or more of the carbon atoms directly acts on less than four other atoms or radicles.[1] Many of the isomeric bodies of this class, while on the whole resembling each other chemically, yet in addition to purely physical differences manifest also minor chemical contrasts. It is of especial significance that the physical differences alluded to do not, as in the case of saturated bodies, involve optical activity.

---

[1] The carbon atoms in ethylene and acetylene are often said to be doubly and trebly linked respectively, and the two substances are represented as follows—

$$
\begin{array}{cc}
\mathrm{H - C - H} & \mathrm{C - H} \\
\| & \| \\
\mathrm{H - C - H} & \mathrm{C - H}
\end{array}
$$

If $\begin{array}{c}\mathrm{C}\\ |\\ \mathrm{C}\end{array}$ symbolises a certain interaction between a pair of carbon

atoms, $\begin{array}{c}\mathrm{C}\\ \|\\ \mathrm{C}\end{array}$ and $\begin{array}{c}\mathrm{C}\\ \|\|\\ \mathrm{C}\end{array}$ certainly suggest interactions of double and treble intensity ; but Thomsen's thermochemical studies in this direction, as well as the salient chemical characteristics of ethylene and acetylene, are at variance with such a state of things. That all the four points of maximal attraction (" bonds ") of the carbon atom must always be in active operation is an unwarranted assumption passed down from the early school of constant valency. The intra-molecular movements of ethylene may be of such a nature that only the three greatest maxima of attraction are able to effectually assert themselves.

Moreover, these cases of isomerism, unlike the optical isomerism just discussed, are not inconsistent with the two-dimensional expression of the doctrine of valency. As illustrative, let us take the case of the isomerism of maleic and fumaric acids.

These two acids differ in their physical properties as follows.   While fumaric acid sublimes on being heated, maleic acid has a definite melting point, 130°; and the latter acid is much more soluble than the former.

While the chief reactions [1] of the two acids demand in each case the same rational formula, *viz.*—

$$\left\{ \begin{array}{l} C.H.COOH \\ C.H.COOH \end{array} \right.$$

yet the following minor differences between the two acids may be cited.   Fumaric acid is more stable than maleic, so that reactions which take place with maleic acid, under ordinary conditions, require high temperatures and high pressures in the case of fumaric acid. Some reactions, *e.g.*, etherification, proceed in the case of both acids under the same conditions, but at a slower rate with fumaric than with maleic acid.   Maleic acid readily yields an anhydride, fumaric acid does not.   The action of bromine on the two acids gives rise to different products; from fumaric acid dibromosuccinic acid results; from maleic, iso-dibromosuccinic acid.   Another peculiarity of these isomers is the readiness with which fumaric acid can be changed into maleic acid derivatives,

---

[1] Both acids are dibasic, and are formed from malic acid by dehydration.   Reducing agents convert both acids into succinic acid.

and maleic acid into fumaric acid derivatives. Thus, if maleic acid is treated with bromine, and then the elements of hydrobromic acid are subsequently removed by the action of water, the result is bromofumaric acid. Conversely, bromomaleic acid can be obtained by a similar series of operations from fumaric acid.

Enough has been said to indicate that we have here a much more pronounced kind of isomerism than that which the lactic acids presented. Moreover, in this case different di-dimensional arrangements of the atoms in the complex $C_2H_2(COOH)_2$ are consistent with the tetravalency of the carbon atom. Hence numerous attempts have been made to assign to the two acids di-dimensional structural formulæ. Such structural formulæ necessarily involve different radicles in the two cases, despite the great chemical similarity of the acids. Thus, Anschütz, while retaining the formulæ

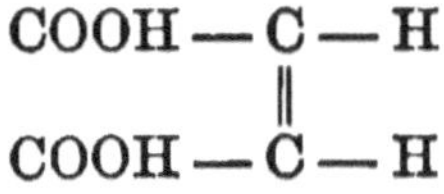

$$\begin{array}{c} COOH - C - H \\ \parallel \\ COOH - C - H \end{array}$$

for fumaric acid, advocates

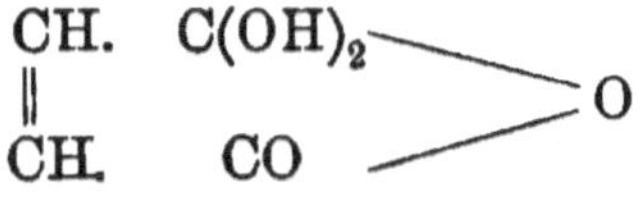

$$\begin{array}{cc} CH. & C(OH)_2 \\ \parallel & \qquad\qquad > O \\ CH. & CO \end{array}$$

as best representing the molecular structure of maleic acid. But these di-dimensional representations met

M

with so little favour generally that the term allo isomerism was provisionally introduced to group together such apparently inexplicable cases of isomerism among unsaturated bodies as are typified by the acids under discussion.[1]

When recourse is had to spatial considerations we find that all difficulty disappears, and that the retention of the term allo-isomerism is needless.   For there are two, and only two, tri-dimensional arrangements

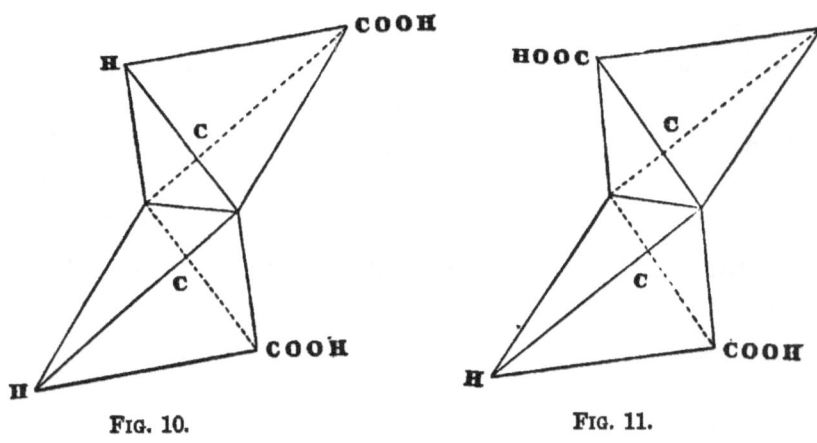

FIG. 10.   FIG. 11.

of the atom complex $C_2H_2(COOH)_2$, in each of which essentially the same structural units obtain, i.e., the same radicles are involved in each case.   These two arrangements will be sufficiently obvious from the accompanying plane projections.

So far our conceptions of valency have been purely

[1] There are, however, many who still maintain that the di-dimensional formulæ represent the properties and peculiarities of the two acids better than do the accepted tri-dimensional ones.

statical, but Wislicenus, by introducing dynamical ideas, has much extended the original theory of van't Hoff and Le Bel. According to Wislicenus, the atoms in a molecule exert influences on each other even when, according to the teachings of valency, they do not *directly* interact, *i.e.*, link each other. The negative or chlorous atoms have a great attraction or affinity for the positive or basylous atoms, and this attraction is either partially satisfied in the molecule by the chlorous atoms swinging themselves as near as possible to the basylous atoms; or it may be that the attraction only sets up an intra-molecular stress which, under favourable conditions of temperature, &c., asserts itself, and causes such a rotational movement in the molecule that the mean distance between the mutually attracting atoms is made as small as possible.

In terms of these views, not only is it possible in many cases to assign the appropriate formula to a given isomer, but obscure chemical transformations, such as the before-mentioned mutual convertibility of fumaric and maleic acid derivatives, find a full explanation. Indeed Wislicenus' views have raised the van't Hoff-Le Bell hypothesis from the level of mere co-ordination and explanation of known facts, to the higher level of prophecy; it now not only explains, but anticipates facts. For a complete exposition of Wislicenus' views we would refer the reader to the pamphlet *Uber die räumliche Anordnung der Atome in organischen Molekulen* (Hirzel, Leipzig).

All we can do here is to exemplify these new views very briefly as they bear on the isomerism of maleic and fumaric acids.

The first question to be settled is—Of the two formulæ, which is to be assigned to fumaric, which

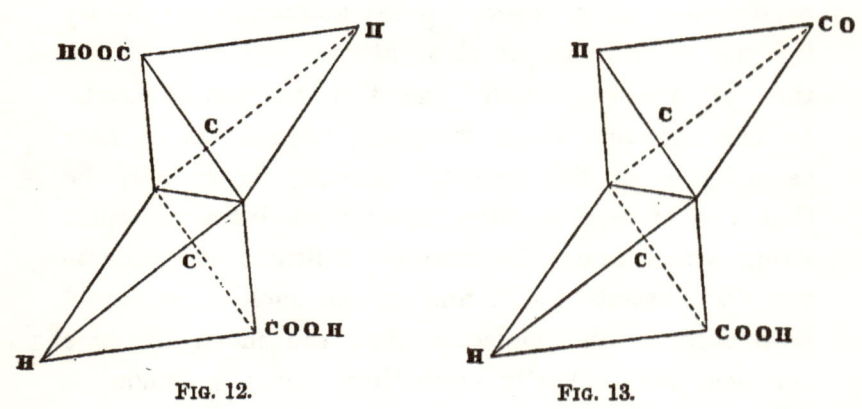

FIG. 12.   FIG. 13.

to maleic acid? The left-hand formula figures a more stable system than does the right-hand one, for in the former the chlorous carboxyl groups are as near as is possible to the basylous hydrogen atoms. But fumaric acid is, as we have said, a more stable acid than maleic; hence the left-hand formula symbolises fumaric acid, and, by exclusion, the right-hand formula maleic acid. The constitution of maleic acid explains its ready dehydration. The water which splits off during the formation of an anhydride from an acid is known to result from the hydroxyl portions of carboxyl groups, and it is but natural to assume that the proximity in the molecule of these groups would favour a reaction in which both are simultaneously implicated.

Let us now explain by a typical example the *rationale* of the formation of maleic acid derivatives from fumaric acid. When bromine and fumaric acid are heated together, each molecule of the acid takes up two atoms of halogen. This, in terms of stereometric formulæ, can only take place in one way :—

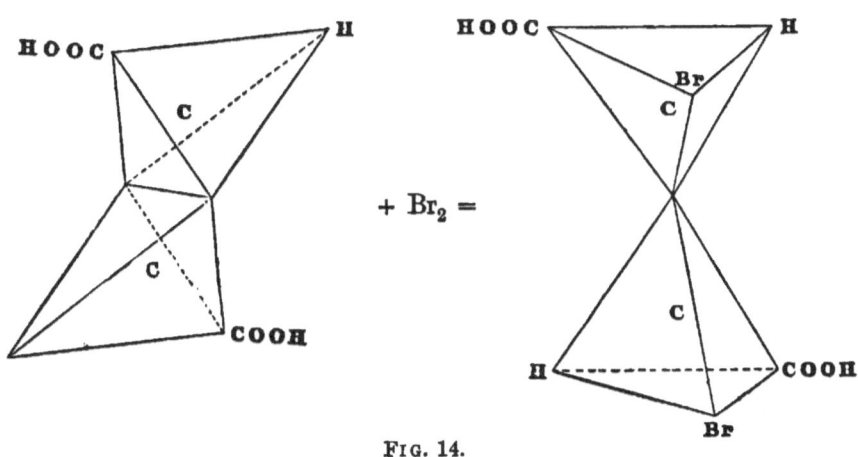

FIG. 14.

But the resulting molecule pictured is in a state of internal stress by reason of the attractions of the strongly positive hydrogen atoms for the strongly negative bromine atoms. As a result of this stress, an intra-molecular rotation around the axis joining the two asymmetric carbon atoms ensues, so as to bring the hydrogen atoms into as close proximity as possible to the bromine atoms.[1]   Hence, when by the subsequent action of water a molecule of hydrobromic acid

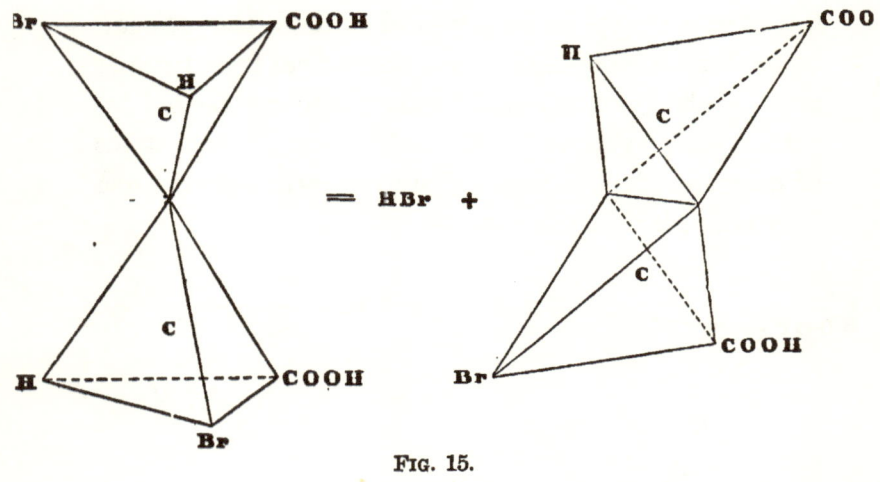

Fig. 15.

is removed, the only possible result is bromo-maleic acid.

[1] From the phenomena presented by the isomerism of the three benzil dioximes, all of which have the formula
$$\begin{array}{c} HON.C.C_6H_5 \\ | \\ HON.C.C_6H_5 \end{array}$$
Meyer and Auwers concluded that carbon atoms may be linked in two ways ; one way admitting of free rotation of the parts of the molecule round the joining axis, and another way inconsistent with any rotational movement.   It has, however, been pointed out, that the exceptional isomerisms exhibited by the benzil dioximes might be explained without having recourse to the idea of non-rotational, singly linked, carbon atoms, by applying stereometric considerations to the nitrogen atom—*e.g.*, regarding it as occupying one angle of a tetrahedron, its three "valencies" or "bonds" being directed towards the other three angles.   But as yet nothing very definite can be said concerning the stereochemistry of the nitrogen atom ; the whole subject has not yet emerged from the purely tentative stage.   It may here be stated that an attempt has quite recently been made to account for the isomerism exhibited by amido-platinum compounds of the type $PtX_4(NH_3)_2$ (see

Many substances are known, to each of which it seems necessary to assign more than one formula. Such substances therefore exhibit a peculiar kind of isomerism which has been called tautomerism. Some of the reactions of such a tautomeric body suggest one structure for it, while other of its reactions seem to demand quite another structure. It has been proposed to call the alternative structures of a tautomeric body its desmotropic forms or states.

Laar, who first drew attention to this form of isomerism, is inclined to think that the molecular architecture of a tautomeric body is changing from moment to moment—that the structure of hydrocyanic acid is at one instant

$$HNC$$

and at the next instant

$$NCH.$$

Others, however, are of the opinion that one of the structures, being a more stable configuration than the other, really represents the body in the free state, but that this stable form is under certain conditions, and by the action of certain reagents, primarily transformed into the less stable, pseudo, or *labile* form which then undergoes further change. Thus phloroglucone behaves sometimes as a phenol giving metallic derivatives and methyl ethers, at other times as a

note, p. 158), by applying spatial considerations to the platinum atom. The latter is regarded as occupying the centre of a regular octahedron, at the six solid angles of which are placed the atoms and radicles constituting the molecule.

ketone or carboxyl compound giving oximes. The formula appertaining to its stable phenolic form, *i.e.*, to phloroglucone properly so-called, is

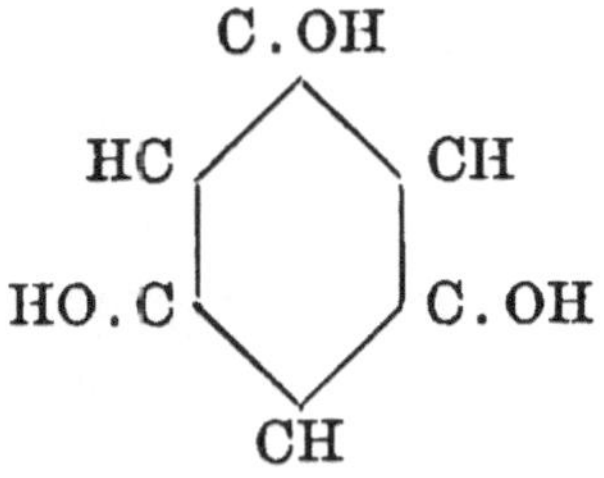

$$
\begin{array}{c}
\text{C.OH} \\
\text{HC} \diagup \qquad \diagdown \text{CH} \\
\text{HO.C} \diagdown \qquad \diagup \text{C.OH} \\
\text{CH}
\end{array}
$$

while its ephemeral *labile* ketonic form demands the structure

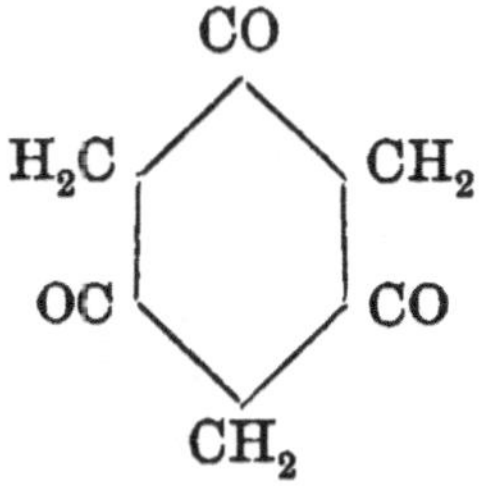

$$
\begin{array}{c}
\text{CO} \\
\text{H}_2\text{C} \diagup \qquad \diagdown \text{CH}_2 \\
\text{OC} \diagdown \qquad \diagup \text{CO} \\
\text{CH}_2
\end{array}
$$

In some instances, both desmotropic forms are capable of continued and well differentiated existence in the free state and even in solution. Thus succinosuccinic ether exists in both a colourless and a yellow modification.

It should be remarked that, so far as is known, all cases of tautomerism depend on the mobility in the molecule of hydrogen atoms; the passage from one desmotropic form to another being effected by the wandering of one or more hydrogen atoms.

These new views may perhaps be adduced to account

for the curious transformations exhibited by cyanogen compounds. Potassium cyanate NCOK prepared from cyanic acid NCOH, when treated with ethyl iodide C_2H_5I gives ethyl isocyanate $OCNC_2H_5$, and not ethyl cyanate as would naturally be expected. Yet isocyanic acid OCNH, the desmotropic form of cyanic acid NCOH, is not known in the free state; the ethyl radicle is required to confer stability on the apparently extremely *labile* configuration OCNH.

Similar considerations may be applied in explanation of the behaviour of cyanuric and thiocyanic acids.

CHAPTER VII.

CHEMICAL EQUILIBRIUM.[1]

ALTHOUGH the occurrence of chemical interaction between solutions of sodium sulphate and hydrochloric acid is not manifest to the unaided senses by reason of the fact that all the factors and products of the interaction are soluble colourless bodies, yet we have convincing indirect proofs that such an interaction does actually take place.[2] Moreover, the evidence in favour of the occurrence of chemical change between solutions of sodium sulphate and hydrochloric acid is of such a nature that we are forced to conclude that the interaction is not correctly represented by the equation

$$Na_2SO_4Aq + 2HClAq = 2NaCl\ Aq + H_2SO_4Aq.$$

For, by convention, this equation when interpreted into words would run:—Dilute solutions of sodium

---

[1] As dissociation phenomena are generally fully treated in elementary text-books, this chapter has been almost wholly devoted to a study of those equilibria which are established in solutions of chemically interacting substances at ordinary temperatures.

[2] See note, p. 193. It is very easy to prove the occurrence of chemical change in the particular system $MgSO_4Aq + 2NaClAq$, although the interaction is not accompanied by visible changes under ordinary conditions. It is only necessary to cool a mixture of the solutions of the two salts, when the very slightly soluble sodium sulphate will partially crystallise out in a hydrated form.

sulphate and hydrochloric acid when brought together in equivalent quantities [1] *completely* decompose each other with the formation of equivalent quantities of sodium chloride and sulphuric acid which both remain in solution.

Now, as a matter of fact, such a complete decomposition of the sodium sulphate and hydrochloric acid does not take place under the specified conditions (dilute solution, and equivalent quantities of the reagents). Only a portion of the system $Na_2SO_4 + 2HCl + Aq$ is changed into the system $2NaCl + H_2SO_4 + Aq$. This fact might be adequately represented in the following manner—

$$(m + n)\ Na_2SO_4Aq + (m + n)\ H_2Cl_2Aq = m\ Na_2Cl_2Aq + m\ H_2SO_4Aq + n\ Na_2SO_4Aq + n\ H_2Cl_2Aq$$

From this representation we learn that when $(m + n)$

[1] Those quantities of acids which neutralise a fixed quantity of any base are equivalent. Conversely those quantities of bases which neutralise a fixed quantity of any acid are equivalent. This was the original and most obvious meaning of equivalency, but the original meaning of the term has been widened so that it now applies to salts as well as to acids and bases. A quantity, x grs., of a salt (MA) is said to be equivalent to a quantity y grs. of an acid A′, if the quantity of acid A necessary to produce x grs. of the salt MA is equivalent in the original acceptation of the word to y grs. of the acid A′. Thus the molecular weight of sodium sulphate interpreted in grams [142 grs.] is formed from the molecular weight in grams [98 grs.] of sulphuric acid. But 98 grs. of sulphuric acid is equivalent to, *i.e.*, will neutralise the same quantity of any base as, 73 grs. of hydrochloric acid or two molecular weights of hydrochloric acid expressed in grams. Therefore in accordance with the widened significance of the term equivalence now in vogue, $2HCl$, H_2SO_4, Na_2SO_4 represent equivalent quantities of the substances formulated no matter what mass units be used.

equivalents of sodium sulphate are brought together with an equal number of equivalents of hydrochloric acid in dilute aqueous solution, only m equivalents of the two substances undergo chemical change, the residual n equivalents of each substance remaining unchanged. But this is not a perfectly satisfactory method of representation, for this reason. It suggests that the molecules of each of the reacting substances must differ in some way or other among themselves, seeing that some enter into chemical reaction under the conditions, while others apparently do not. And this is not in conformity with the fundamental principle of the atomic-molecular theory, which principle asserts that the molecules of a substance are all precise replicas one of another, both in structure and attributes. If one molecule of sodium sulphate reacts under certain conditions, then we should expect *all* the molecules of sodium sulphate to react under these conditions.

The difficulty disappears at once if we regard reactions like the one under consideration as the result of two independent and antagonistic chemical changes proceeding simultaneously within the system.[1]

When hydrochloric acid is added to an equivalent quantity of sodium sulphate in dilute aqueous solution,

[1] That two or more reactions can proceed within a system simultaneously and independently one of another is the enunciation of the principle of the coexistence of reactions—a principle which bears much the same relation to chemistry as does Newton's second law of motion to dynamics. This principle has never been *directly* proved; but reasoning founded on the assumption of such mutual independence of reactions has led to results in harmony with experimental fact, and thus the truth of the principle is established.

molecules of sodium chloride and sulphuric acid are formed at a gradually diminishing velocity as time goes on.[1] The new molecules thus formed do not, however, remain inert, but interact to reproduce molecules of the original substances at a gradually increasing velocity. Let us call the passage from $Na_2SO_4Aq + H_2Cl_2Aq$ to $Na_2Cl_2Aq + H_2SO_4Aq$ the *direct partial change*, and the passage in the opposite direction the *reverse partial change*.

What causes the velocities of the two changes thus defined to vary from moment to moment? Guldberg and Waage first definitely answered this question, by asserting that the velocity of every change at any moment varies directly as the product of the number of the equivalents of the factors of the change present in unit volume of the medium of change. As this number is continually varying, owing to the occurrence of change, it follows that the velocity must also continually vary.

---

[1] No chemical change is instantaneous, but every change requires for its occurrence a period of time varying from a fraction of a second too small to measure, to several years. The reactions between various acids and alkalis (*i.e.*, neutralisation) take place in so short a time that it has not been possible to measure their absolute durations. Estimates of their relative durations have, however, been indirectly obtained. On the other hand, many of the reactions between organic bodies proceed very slowly indeed. Witness in this connection the slowness of the chemical changes occurring in the so-called ageing of wine. The use of the word velocity in connection with chemical reactions as introduced by Wenzel in 1777 demands explanation. The original dynamical meaning of the term is space passed over per unit time. But the idea of space does not enter into the definition of the term in its chemical acceptation. The velocity of a chemical reaction at any moment simply denotes the quantity of substance (expressed in equivalents) that undergoes change per unit of time at that moment.

Suppose the two substances A and B combine completely to form a third body C.   Let us bring together solutions of A and B, such that $2p$ equivalents of A and $2q$ equivalents of B are contained in unit volume of the resulting solution.   Then the velocity, V, of the reaction (*i.e.*, the rate at which C is formed) will vary as $2p \times 2q$.   Or stated algebraically—

$$V \propto 4pq$$
$$V = 4Kpq.$$

Where K is a constant called the *velocity constant* for the change.[1]   After a certain time there will be left only p equivalents of A, and q equivalents of B in unit volume of the medium of change.   The velocity of formation of C at this epoch, according to Guldberg and Waage, would be Kpq—only ¼th its original value. The principle here illustrated is often referred to as Guldberg and Waage's law of mass action.[2]

To return now to the interaction of hydrochloric acid and sodium sulphate.   The number of equivalents

[1] The *velocity constant* K must be carefully distinguished from the *velocity* V.   Under given conditions K is constant for any given change, but V of course varies from moment to moment.

[2] The first to prove the influence of mass in chemical changes was Wenzel, 1777.   Wenzel, however, confined his investigations to a single class of changes, viz., the interaction of metals with acids of various strengths.   In the beginning of this century, Berthollet asserted the influence of mass in chemical changes in general terms.   "Every chemical change is essentially of the nature of a combination, and the power of any substance to enter into combination is proportional to its affinity *and to its mass*."   This is practically the text of Berthollet's famous work, *Essai de Statique Chimique*.   According to Berthollet, a small affinity can be compensated by a large mass, and hence complete reactions are the exception and not the rule in chemistry.   Suppose

expressed in grams or otherwise (*i.e.*, the masses) of NaCl and H_2SO_4 resulting from the direct change depends not only on the velocity constant peculiar to the change, but also on the numbers of the equivalents (*i.e.*, the masses) of HCl and Na_2SO_4 in unit volume of the solution. As these latter numbers are continually decreasing, the velocity of the direct partial change must continually decrease from a maximum value downwards. But as the numbers of the equivalents per unit volume of the factors of the direct partial change decrease, the numbers of the equivalents of the factors of the inverse partial change *pari passu* increase, consequently the velocity of this change is continually increasing from zero upwards. Hence it necessarily follows that a state of what has been called mobile equilibrium between the direct and reverse partial changes must finally be reached, whereat the velocities of the two changes being exactly equal,

the body C has a stronger affinity for A than has B. Then if C be added to AB in equivalent quantity, the reaction

$$AB + C = AC + B$$

will proceed, but not to completion, for the gradually increasing mass of B will eventually compensate its smaller affinity, and make it formidable enough to contest successfully the diminishing quantities of C. Before Berthollet's time it had been held that reactions are determined entirely by the stronger affinities ; to which view the idea of incomplete reactions was quite foreign. From his fundamental premiss, however, Berthollet made deductions which Proust proved to be incorrect (see p. 74), and this overthrow of the incorrect deductions unfortunately brought the correct premiss from which they were derived into disrepute and temporary oblivion. Guldberg and Waage's law of mass action is nothing else than Berthollet's views reinstated and thrown into mathematically exact form.

effective chemical change one way or the other ceases. When this permanent equilibrium point is arrived at, as many equivalents of HCl and Na_2SO_4 are decomposed by the direct partial change per unit time as are formed by the reverse partial change in the same time unit; but the molecules of Na_2SO_4, HCl, ~~HNO_3~~, and NaCl existing at one epoch are not necessarily the *same* ones which will be present at another subsequent epoch. If they do chance to be identically the same molecules at the two epochs, we can assert that they have not had a continued existence in the interim.

In the representation of these so-called incomplete reactions, wherein a state of equilibrium is finally reached, van't Hoff has proposed to replace the sign of equality in the ordinary equations by a pair of oppositely directed arrows, thus—

$$H_2Cl_2Aq + Na_2SO_4Aq \rightleftarrows Na_2Cl_2Aq + H_2SO_4Aq.$$

Now, thoroughly appreciating the nature of the dynamical equilibrium which marks the cessation, not of chemical change, but of *effective* chemical change, in these incomplete reactions, we will return to that method of their representation which we first gave, and which for the immediate purpose in hand is more convenient than the one proposed by van't Hoff.

Let us suppose that in the general equation

$$(m + n)\,Na_2SO_4Aq + (m + n)\,H_2Cl_2Aq = m\,Na_2Cl_2Aq + m\,H_2SO_4Aq + n\,Na_2SO_4Aq + n\,H_2Cl_2Aq$$

$m + n$ equals 100. The question arises, what are the

individual values of m and n? This question has been submitted to experiment, and it has been ascertained that m and n in the above reaction have the approximate values $66\frac{2}{3}$ and $33\frac{1}{3}$ respectively.[1] Moreover, exactly the same state of final equilibrium is arrived at if, instead of setting out with hydrochloric acid and sodium sulphate, we start with hydrochloric acid and the proximate constituents of sodium sulphate, viz., caustic soda and sulphuric acid, in accordance with the following equation—

$$(m + n)\ \mathrm{Na_2O_2H_2Aq} + (m + n)\ \mathrm{H_2SO_4Aq} + (m + n)$$
$$\mathrm{H_2Cl_2Aq} = m\ \mathrm{Na_2Cl_2Aq} + m\ \mathrm{H_2SO_4Aq} + n\ \mathrm{Na_2SO_4Aq}$$
$$+ n\ \mathrm{H_2Cl_2Aq}.$$

This latter method of regarding the reacting system emphasises the fact that the reaction virtually consists in a strife between two equivalent quantities of acids to appropriate a quantity of base only just sufficient to neutralise either acid separately. As a result of this competition the base divides itself between the two acids, twice as much base (*i.e.*, $66\frac{2}{3}$ equivalents) combining with the hydrochloric acid as combines with the sulphuric acid ($33\frac{1}{3}$ equivalents).

In accordance with the results of such investigations as the one just described, acids are classed as weak or strong. When two acids in equivalent quantities and

[1] For the actual carrying out of methods suited to the investigation of equilibria see Muir and Carnegie's *Practical Chemistry*, pp. 141–146, 176–184.

in dilute aqueous solution are presented to a dilute aqueous solution of a quantity of base just sufficient to neutralise either acid separately (or what amounts to the same thing, when an acid is allowed to react with an equivalent quantity of the salt of another acid in dilute aqueous solution) the base is in general divided between the two acids, and the strengths (or affinities, or avidities) of the two acids are proportionate to the quantities of base each appropriates. Thus it follows that sulphuric acid is a weaker acid than hydrochloric, the latter acid being approximately twice as strong as the former. By conducting numerous quantitative experiments, involving different pairs of acids, we can obtain a table of the relative affinities or strengths of the acids, such as the following one, in which, however, the values assigned the various acids are admittedly only approximate :—

Nitric acid	1·00
Hydrochloric acid	1·00
Hydrobromic acid	·89
Hydriodic acid	·70
Sulphuric acid	·49
Selenic acid	·45
Trichloracetic acid	·36
Orthophosphoric acid	·25
Oxalic acid	·24
Hydrofluoric acid	·05
Citric acid	·05
Acetic acid	·03
Boric acid	·01
Silicic acid	·00

As will be seen nitric and hydrochloric acids are the two strongest acids, being approximately equal in strength. Compared with either of these, acetic acid is seen to be a very feeble one. If equivalent quantities of hydrochloric acid and acetic acid were added to a dilute aqueous solution of caustic soda, containing just enough caustic soda to neutralise either acid separately, the caustic soda would divide itself between the hydrochloric and acetic acids in the ratio of 1 : 03, *i.e.*, about 2·91 per cent. of the soda would combine with the acetic acid and 97·09 per cent. with the hydrochloric acid. In a precisely similar way bases have been classed as weak or strong. Of a pair of bases presented in dilute aqueous solution to an acid only just sufficient to neutralise either base separately, that base which appropriates, or combines with, the greater part of the acid is the stronger base. The following table gives the relative strengths of a few of the better investigated bases in terms of the arbitrary value unity assigned to the strongest base, lithium hydroxide :—

Lithium hydroxide	1·00
Sodium hydroxide	·98
Potassium hydroxide	·97
Thallium hydroxide	·90
Piperidin	·16
Ammonia	·02

We have seen that hydrochloric acid is approximately twice as strong as sulphuric acid if the criterion be a tug-of-war for a quantity of soda insufficient to meet the conjoint demands of the two acids. It has been found that approximately the same relation holds between the strengths of the two acids whatever base be made the object of competition, whether potash, or soda, or lime, &c. Moreover, as is well known, acids have the power of inverting cane sugar, that is, of transforming it in the presence of water into a mixture of dextrose and lævulose ; and also in this transformation hydrochloric acid is found to act twice as energetically as sulphuric acid. Numerous other reactions might be adduced which are either brought about or accelerated by acids, and in all of them hydrochloric acid is approximately twice as effective as sulphuric acid.

Similar remarks apply to changes induced or accelerated by bases. If lithium hydroxide is fifty times as strong as ammonia when the struggle for a particular acid is the criterion, then it is found to be approximately fifty times as effective as ammonia in bringing about any other change demanding the intervention of a basic substance.

On the basis of the approximately concordant results for the strengths of acids and bases derived from the investigation of such diverse changes as we have hinted, it was at one time claimed that the values thus assigned to the acids and bases as representing their strengths, could they be freed from all experimental error, would be perfectly characteristic numbers for the respective acids and bases, quantitatively conditioning all the reactions brought about by them. In pursuance of these views, it was proposed to give the generic name of *specific affinity constants* to the numbers representing the strengths of the acids and bases.

On fuller investigation, however, it appeared that neither the claim nor the proposal could be admitted; for the partition of a base between two acids depends not only on the nature of the contending acids, but also on the quantity of water present. A table of strengths or affinities constructed from the results of experiments on the partition of a base between acids in normal solution,[1] does not present the acids in the same order as a table founded on the results of investigations of deci-normal acid solutions. In other words, water cannot be regarded as a merely passive medium when the power of acids is in question; "the power of an acid to do" is a function of the state of dilution of the acid. For instance, in the system *acid A, acid A', base, water*, we have not merely a tug-of-war of the two acids A and A' for the base, but the final equilibrium is the expression of the resultant of (1) the "affinity" of A for water, (2) the "affinity" of A' for water, (3) the "affinity" of the base for water, (4) the "affinity" of A for the base, and (5) the "affinity" of A' for the base; and in con-

---

[1] Let the numerical value of the quotient—

$$\frac{\text{molecular weight of acid}}{\text{number of replaceable H atoms in molecule}}$$

be called n. Then a normal solution of an acid contains n grams of the pure acid per litre of solution at a specified temperature. A deci-normal solution has $\frac{1}{10}$th the concentration of a normal one.

formity with the law of mass action, every alteration
in the quantity of water present necessarily alters
the final distribution of matter marking the state of
equilibrium from which the strengths of the acids are
judged.

Quite recently, however, it has been found possible
to obtain for the acids characteristic numbers which
are independent of their greater or less dilution, and
of the particular reactions in which the acids may
be implicated. These numbers are now regarded as
the true affinity constants of the acids, but they
necessarily have quite a different significance from the
old system of supposed constants derived from the
study of the partition of acids between bases, &c.,
seeing that they are independent of dilution.

When an acid is dissolved in a large quantity of
water, it behaves in many respects as if it were (and
is accordingly by some believed to be) partially decomposed into its ions (see note, p. 153). Thus a dilute
solution of hydrochloric acid behaves as if it contained,
in addition to molecules of HCl, ions of hydrogen and
chlorine. If at a certain dilution K the number of
undecomposed molecules is just equal to the number
decomposed into ions, then K is taken as a measure of
the true specific affinity constant of the acid.

The investigation of affinity in accordance with these
conceptions has as yet been chiefly confined to organic
acids, and it is found that the *order* of the acids in a
table of affinities, according to the new definition of
affinity, is practically the same as the *average order* in

the old tables of supposed affinity constants which were derived from the results of different reactions brought about by acids of varying concentration.[1]

We have seen (page 57) that certain phenomena exhibited by fairly strong aqueous solutions point to definite combinations between water molecules and molecules of dissolved substance, $i.e.$, to the existence in solutions of definite liquid hydrates. It is supposed that these liquid hydrates are very unstable bodies which at ordinary temperatures are partially dissociated,[2] the original hydrates forming with the products of their dissociations mobile equilibria as represented in the following equations :—

$$m\,(An\,H_2O) \rightleftarrows mn\,H_2O + m\,A$$
$$m'\,(An'\,H_2O) \rightleftarrows m'n'\,H_2O + m'\,A \ \&c.,$$

where $(m + m' + \&c.)$ A represents the salt dissolved in $mn = m'n' = \&c.$ molecules of water. The interpretation of the above symbolism is that in a solution of A of determined concentration, the hydrates AnH_2O, $An'H_2O$ &c. are present, but that these hydrates

---

[1] In the algebraical expression through which the desired affinity constants are obtained from the results of experiment, there is an inherent weakness which makes itself especially felt in the case of the stronger inorganic acids. For further details on the subject of affinity constants the reader is referred to the *Lehrbuch der Allgemeinen Chemie* of Ostwald, to whom we owe much of the best work that has been done in the direction of raising affinity to a quantitative dignity.

[2] A dissociation is a reversible decomposition. When $KClO_3$ is heated it *decomposes* into KCl and 3O, for these two substances do not recombine on cooling to reform potassium chlorate. When PCl_5 is heated, it *dissociates* into PCl_3 and Cl_2, because these two gases as they cool in contact combine together completely to reform PCl_5.

partially dissociate into anhydrous salt,[1] (mA, $m'A$ &c.) and water (mnH$_2$O, $m'n'$H$_2$O &c.) till mobile equilibria are established for each system, *i.e.*, until in each system as many molecules of the hydrate are dissociated as are reformed per unit of time. As the concentration of the solution is varied, the ratios in which the various dissociating hydrates are present also varies, and at certain stages in the concentration new dissociating hydrate systems would make their appearance, while pre-existing systems would disappear. According to the views here illustrated, solutions have been defined as "fluid and unstable, but definite chemical compounds in a state of dissociation."

All the incomplete reactions we have already considered have been homogeneous,[2] but homogeneity in

---

[1] Or it may be, into a hydrate of lower hydration in accordance with the equation

$$m\,(A\,n\,H_2O) \underset{\rightarrow}{\leftarrow} m\left(A\frac{n}{x}H_2O\right) + \left(mn - \frac{mn}{x}\right)H_2O,$$

where m, n, and x are integers.

[2] When all the products and factors of a reaction are in the same physical state (same state of aggregation), the reaction is said to be homogeneous. The following are examples of homogeneous reactions :—

Cl$_2$ + H$_2$ = 2HCl.   NaCl Aq + KNO$_3$Aq $\underset{\rightarrow}{\leftarrow}$ KCl Aq + NaNO$_3$ Aq.

$\underbrace{\qquad\qquad}$ $\underbrace{\qquad\qquad\qquad\qquad\qquad\qquad}$

gases.   solutions.

Fe + S = FeS.

$\underbrace{\qquad\qquad}$

solids.

The combination of gaseous SO$_2$ with solid PbO$_2$ to form solid PbSO$_4$ is illustrative of non-homogeneous reactions. The significance of the terms homogeneous and non-homogeneous used in connection with reactions or systems must be carefully distinguished from the significance of the same terms when used, as in Chap. III., in connection with substances.

this sense is not an essential condition of incomplete reaction. The very familiar reaction between solutions of common salt and silver nitrate is not correctly represented in the equation

$$AgNO_3Aq + NaCl\,Aq = AgCl + NaNO_3Aq,$$

for when effective chemical change between strictly equivalent quantities of $AgNO_3$ and $NaCl$ has ceased, it is easy to prove the presence in the non-homogeneous system of traces of undecomposed $AgNO_3$ and $NaCl$. In other words, a condition of equilibrium finally supervenes, the equilibrium in this case, however, being so much in favour of the direct partial change, that unless for very accurate work, the decomposition of $AgNO_3$ by an equivalent quantity of $NaCl$ may be regarded as practically complete.

The interaction of steam and heated iron is another familiar instance of an incomplete reaction taking place in a non-homogeneous system.

$$3Fe + 4H_2O \leftrightarrows Fe_3O_4 + 4H_2$$
$$\text{(solid)} \quad \text{(gas)} \quad \text{(solid)} \quad \text{(gas)}$$

The sole condition for incompleteness of reaction between equivalent quantities of interacting substances resulting in mobile equilibria is that the products of the change must be of such a nature that they are all retained within the sphere of action of the system considered.[1] If by any means removal from the sphere

[1] It follows that the salts in solution in mineral waters, &c., must constitute very complicated dynamic equilibria. The schemes in which analysts are wont to express the results of their analyses are mislead-

of action of one or more of the products of the change is effected, the reaction proceeds to completion. Thus, if in the interaction of steam and heated iron it were possible to allow the hydrogen to escape as soon as formed, without at the same time permitting any steam to escape, then three atoms of iron would completely decompose four molecules of steam. One of the factors necessary to the reverse partial change being removed, that change cannot take place, and the direct partial change meeting with no opposition proceeds to completion. To take another instance actually realised by Berthelot and St. Gilles. When benzoic acid and alcohol are brought together they slowly interact with the production of benzoic ether (ethyl benzoate) and water, which in turn simultaneously interact, reforming the original factors of the change.

$$C_6H_5COOH + C_2H_5OH \leftrightarrows C_6H_5COOC_2H_5 + H_2O.$$

Under ordinary conditions effective chemical change ceases when about 66 per cent. of the benzoic acid and alcohol have been transformed into benzoic ether. But when the water resulting from the change is removed from the sphere of action, the whole of the benzoic acid and alcohol is forthwith etherified. This removal of the water from the sphere of action is effected by the addition of barium oxide to the system. As soon as

ing. The bases in a homogeneous system can be accurately determined and also the acids, but it is impossible from the results of the quantitative analysis only to apportion the acids to the bases so as to represent the actually existing constitution of the system.

any water is formed it combines with the oxide forming inert barium hydroxide.

It is to be noted that the same state of equilibrium is reached whether we set out with the system $C_6H_5COOH + C_2H_5OH$ or with the system $C_6H_5COOC_2H_5 + H_2O$. In the former case effective change proceeds till 66 per cent. of the benzoic acid and alcohol have been transformed into benzoic ether and water; in the latter case effective change proceeds until 66 per cent. of the ether and water remain undecomposed. This is a perfectly general characteristic of these mobile equilibria. Representing incomplete reactions by the general equation

$$\underbrace{AB + A'B'}_{(1)} \leftrightarrows \underbrace{A'B + AB'}_{(2)}$$

we may say that in all cases the distribution of matter which obtains when equilibrium is established is independent of whether the point of departure be made the system (1) or (2). In a few particular cases this statement admits of very simple proof. For instance, a mixture of equivalent quantities of KClAq and $NaNO_3Aq$ gives exactly the diffusate as a mixture of equivalent quantities of KNO_3Aq and NaClAq, showing that in both cases the *same* equilibriated system

$$x\,NaClAq + x\,KNO_3Aq + (1 - x)\,KClAq + (1 - x)\,NaNO_3Aq$$

is undergoing dialysis.

Another method of rendering an incomplete reaction practically complete as regards some one or more of the members constituting the system will at once be evident on referring to Guldberg and Waage's law of mass action. This method consists in largely increasing the mass of one of the factors of the change relatively to the other. Thus, if in the etherification of alcohol by benzoic acid we largely increase the number of equivalents of alcohol relatively to the number of equivalents of benzoic acid, or *vice versâ*, then, in either case, we approach indefinitely close to a state of complete etherification of the acid on the one hand, or the alcohol on the other.

It is in virtue of the action of mass that we are able to completely convert $NaCl$ into $NaNO_3$ by means of HNO_3Aq, or conversely to change $NaNO_3$ into $NaCl$ by means of $HClAq$. When $NaCl$, for instance, is treated with HNO_3Aq a partial change takes place, and $NaCl$, HCl, $NaNO_3$, and HNO_3 are all present. If the system be now heated, the free acids owing to their volatilities pass away, leaving non-volatile $NaCl$ mixed with a little non-volatile $NaNO_3$. If more HNO_3Aq is added to this residue, and the system again heated, a second residue richer in $NaNO_3$ will result; and so by repeating the operations of adding HNO_3Aq and evaporating often enough (*i.e.*, by using a large enough relative mass of HNO_3), the $NaCl$ can be completely changed into $NaNO_3$. The case of the converse transformation of $NaNO_3$ into $NaCl$ by means of a large excess of HCl is quite similar.

In the light of the facts just treated, a reference to the remarks made on solution both in this chapter and in Chapter III. leads us into difficulties.

If a substance in solution be largely diluted, *i.e.*, if the relative mass of water be largely increased, we should expect, in accordance with what precedes, that the excess of water would lead to a great increase in the amounts and stabilities of the higher hydrates of the dissolved substance. But those who have experimented most with very dilute solutions, find that their results are best summarised in terms of the hypothesis that hydrates do not exist at all in dilute solutions. It appears as if the increased dilution not only breaks up the molecular complexes of water molecules and salt molecules constituting the hydrates, but actually destroys the integrity of the salt molecules themselves, resolving these into their ions.[1] For the present, until more light is shed on the vexed question of solution, we must be content to remember that this difficulty exists without attempting to decide it one way or the other.

The question now arises for discussion, are all reactions between equivalent quantities of mutually interacting bodies incomplete? May not one or more of the products of certain changes be of such a nature

---

[1] See note, p. 153. An ion may consist of a group of atoms, *e.g.*, SO_4, or of a single atom, *e.g.*, Na. In the latter case the ion and the atom are supposed not to be identical. For the present it is customary to attribute the assumed difference between them to the possession by the ion of a certain charge of electricity proportional to the valency of the atom.

that they are removed from the system's sphere of action as quickly as they are formed, in virtue of their own peculiar properties? Changes giving rise to bodies with such properties, would proceed to completion even although their factors were brought together in strictly equivalent quantities.

Stas has shown that AgBr is not acted on at all by a solution of $NaNO_3$, or that if there be any action it is too slight to be detected by the analytic means at our disposal. Hence it follows that the reaction between $AgNO_3Aq$ and NaBrAq is, so far as we can tell, a complete one, and that the equation—

$$AgNO_3Aq + NaBrAq = AgBr + NaNO_3Aq$$

is experimentally realised.

There are many well-known reactions, such as that between H_2SO_4Aq and $BaCl_2Aq$, $viz.,$

$$H_2SO_4Aq + BaCl_2Aq = BaSO_4 + 2HClAq,$$

which are for practical purposes regarded as complete, but which, as a matter of fact, are incomplete reactions in which the distribution of matter marking the attainment of equilibrium is overwhelmingly in favour of the direct partial change. The $BaSO_4$, formed in the above reaction, is slightly soluble in (*i.e.*, interacts with) dilute solutions of HCl, and, in virtue of this slight solubility, there is a reverse partial change, which is the condition for incompleteness of reaction and the establishment of an equilibrium. But the reverse partial change in this case is of very small moment.

The subject of neutralisation calls for a word or two here. Those particular and important branches of volumetric analysis known as alkalimetry and acidimetry are founded on the assumption of the absolute truth of such equations as—

$$H_2SO_4Aq + 2NH_4OHAq = (NH_4)_2SO_4Aq + 2H_2O$$
$$HClAq + NaOHAq = NaClAq + H_2O.$$

That is to say, these reactions are regarded as complete although there is evidently no removal from the sphere of action, artificial or otherwise, of the products of the neutralisations of equivalent quantities of acids and bases. Here, again, it is probable that we have instances of incomplete reactions in which the reverse partial change is so insignificant as to be negligible. That ammonium sulphate is slightly decomposed by water with the formation of free acid and ammonia is probable from the fact that air when passed through solutions of $(NH_4)_2SO_4$ acquires alkaline properties— the solutions themselves meanwhile developing acid characters. This behaviour of ammonium sulphate solution has been adduced to explain the completeness of the interaction of equivalent quantities of ammonium sulphate $(NH_4)_2SO_4Aq$ and potassium carbonate K_2CO_3 Aq. Seeing that both the factors of this interaction, and all the possible products formed by double decomposition, are soluble in water and therefore remain within the sphere of action, one would be inclined, off-hand, to class this particular change with the incomplete reactions.

According to Berthelot the completeness of the interaction finds an explanation in terms of the partial
decomposition of one of the interacting factors.   The
ammonium sulphate and water may be regarded as
giving rise to some such equilibrium as is represented
in the following equation—

$$(NH_4)_2SO_4Aq + 2H_2OAq \leftrightarrows 2NH_4OIIAq + H_2SO_4Aq.$$

The potassium carbonate destroys this equilibrium by
neutralising (and so removing from the sphere of action)
the free sulphuric acid present at any moment; more
ammonium salt reacts with water in the tendency to
again establish the primitive equilibrium, and so the
cycle repeats itself until practically the whole of the
ammonium sulphate is decomposed.

The experimental investigation of the possibility of
the establishment of equilibria in the cases of reactions
evolving gases (which under ordinary circumstances
escape from the sphere of action) is a very difficult
matter.   A definite answer has as yet not been furnished
to the question, is the equation

$$Zn + H_2SO_4Aq \leftrightarrows ZnSO_4Aq + H_2,$$

a correct representation of facts, provided the hydrogen
evolved is prevented from leaving the system—would
the reaction under such conditions be found to be
actually incomplete?   In the meantime one has little
hesitation in giving a theoretical answer in the affirma-
tive to such a question.

The existence of such equilibria as we have been

discussing disproves a principle which was enunciated by Thomsen in 1854, and again in 1864 by Berthelot. The latter naturalist has done so much work in the attempt to establish the principle on an experimental basis that it is now generally referred to as Berthelot's *law of maximum work*, although priority in the matter undoubtedly belongs to Thomsen.[1] Berthelot enunciated his " law " in the following terms :—

Every chemical change realised without the intervention of external energy tends to the formation of that body or system of bodies, the production of which is accompanied by the development of the greatest quantity of heat.

This statement, which simply asserts that every chemical reaction tends to make the system assume that state, in the attainment of which it liberates most heat, further implied to Berthelot the *necessity* for the occurrence of every transformation that would involve an evolution of heat, and the impossibility of the spontaneous occurrence of every transformation that would involve an absorption of heat.[2] And it is this

---

[1] Berthelot called his principle the " Law of maximum work," from its analogy to a well-established principle in mechanical energetics. If a raised stone be allowed to fall, it falls vertically to the earth's surface, and never in a direction inclined to the plumb line. The vertical direction is the one in which occurs a maximum change of potential into kinetic energy per unit time. This is an instance illustrating the so-called maximum principle which asserts that of all possible changes that one will take place which involves the greatest transformation of energy per unit time. (Throughout the remainder of this chapter it is assumed that the reader has an elementary knowledge of general energetics.)

[2] As a corollary to this " law," Berthelot stated his theorem of the necessity for reactions as follows :—" Every chemical change which

practical aspect of the "law" that has chiefly appealed
to, and engaged the attention of, chemists generally.

Suppose we have a substance quantitatively represented by AB, and we wish to know whether it will
interact with a third body quantitatively represented
by C to form either the bodies AC + B or the bodies
BC + A. We may provisionally regard such an interaction as taking place in two consecutive stages; in
the first stage AB is decomposed into A and B, and
then in the second stage A combines with B or C, as
the case may be. If A and B combine together to
form AB with the evolution of a quantity of heat
energy h, we know that exactly the same quantity of
energy h must be added in order to reverse the change
and resolve AB into A and B.[1] Suppose that the
combination of A and C gives rise to an evolution of
heat energy represented by H, and that the combination of B and C is attended by an evolution of heat H'.

can be accomplished without the aid of a preliminary action or the
addition of energy from without the system, necessarily occurs if it is
accompanied by disengagement of heat." This theorem seems to have
been generally applied without regard to the significance of the somewhat indefinite qualifications "which can be accomplished . . . from
without the system."

[1] The principle, that the quantity of heat energy absorbed in decomposing a given mass of a compound into its elements is exactly
equal to the heat energy evolved when the elements combine to form
the given mass of the compound, was first given by Lavoisier and
Laplace. This principle, which is a necessary consequence of the
more general principle of the conservation of energy, is also true for
chemical changes other than mere synthesis and analysis. If the
passage of a complex system from any state A to any other state B,
by any path whatsoever, absorbs a quantity of heat energy H, then
the reverse passage from B to A by any path will be attended with
the evolution of H heat units.

The following cases must be considered :—

(1.) H and H′ may both be less than h, in which event the "law," or rather its popularised implication, pronounces against the spontaneous occurrence of either change, seeing that both changes under these conditions would be endothermic.[1] (2.) H and H′ may both be greater than h, in which case both reactions are exothermic, and therefore possible under the conditions. The relative magnitudes of H and H′ according to the "law" decide as to which of the two possible changes will actually occur. If H > H′, then AC + B will be formed exclusively ; if H′ > H, then BC + A will be the sole products of the reaction.

When steam is passed over hot iron, chemical change ensues, resulting in the production of magnetic oxide of iron and hydrogen

$$3Fe + 4H_2O = Fe_3O_4 + 4H_2,$$

but if retaining the same conditions copper be substituted for iron, the steam is *not* decomposed with the production of copper oxide and hydrogen in accordance with the equation

$$Cu + H_2O = CuO + H_2.$$

These facts and many others of a similar nature might be adduced as confirmations of Berthelot's "law."

[1] It is to be regretted that the application of so many different significations to the terms exothermic and endothermic has led many chemists to avoid their use altogether. Rigidly defined they are exceedingly convenient terms for the description of thermo-chemical phenomena. In the text an endothermic reaction is taken to mean any reaction which is accompanied by an absorption of heat ; an exothermic reaction, any reaction which is accompanied by an evolution of heat. The terms as thus used have no quantitative significance.

For since the formation of a gram molecule (232 grs.) of Fe_3O_4 from its elements is accompanied by an evolution of 264,700 heat units, while the resolution of four gram molecules (72 grs.) of steam into its elements is attended by the absorption of only 232,000 heat units, it follows that the realisation of the first change will be attended by the evolution of heat energy to the extent of 32,700 units. On the other hand, the 37,200 units of heat evolved when a gram molecule (80 grs.) of copper oxide is formed from its elements is less than that required, 58,000 units, to resolve a gram molecule (18 grs.) of gaseous water into its elements, and so the realisation of the second reaction would involve an absorption of heat.[1]

Further, the generalisation that the readiness and intensity with which reactions take place increase with the thermal values of the reactions, is also in accordance with Berthelot's "law."

But on the other hand there are very many facts which are at variance with the "law." Suppose we wish to know, without actually trying the experiment, whether hydrochloric acid gas acts on silver in accordance with the empirical equation

$$HCl + Ag = AgCl + H.$$

Looking up the *heat of formation* of HCl (*i.e.*, the

---

[1] The possibility that reactions could take place in such way that the hydrogen of the water combines with the metals and the oxygen is set free, is presumed to be precluded by the fact that such changes would be strongly endothermic, though through lack of data the exact values of the heat absorption in each case cannot be given.

heat evolved when a gram molecule of hydrochloric acid gas is formed from its elements), we find it to be 220,000. The heat of formation of AgCl is 294,000. Now, applying Berthelot's "law" to these data, we are led to the conclusion that silver will be attacked by hydrochloric acid. But this conclusion is false, for silver is quite unaffected by hydrochloric acid. Again, when electric sparks are sent through a mixture of 2 volumes of hydrogen, with 2 volumes of chlorine and 1 volume of oxygen, hydrochloric acid is exclusively formed, although the formation of water from the hydrogen and oxygen would be attended by a far greater production of heat.

But the existence of chemical equilibria is, as we have already hinted, the most generalised argument that can be brought forward in opposition to the "law" of Berthelot. Let us suppose that the direct partial change of a particular case of equilibrium is attended by a heat evolution, then it follows from the well-established principle of the conservation of energy that the reverse partial change must involve a heat absorption. For the same reason, if the direct partial change is endothermic instead of exothermic, then the reverse partial change must be exothermic instead of endothermic. That is to say, an equilibrium demands the occurrence of two simultaneous chemical changes, one of which is associated with an evolution of heat, the other with an absorption of heat; and when equilibrium is attained, these two changes are proceeding at exactly the same rates. If Berthelot's "law" were true, all reactions would belong to the complete type, and incomplete reactions, *i.e.*, equilibria, would be impossible.

If we interpret Thomsen and Berthelot's principle as merely asserting that every *purely* chemical change of a complete type that takes place spontaneously must be accompanied by a loss of chemical energy which will finally appear as heat—the lowest form of energy —then we are bound to admit the truth of the principle at the same time as we deprive it of all practical import. For all the so-called chemical changes with which we have to do are not *purely* chemical changes, in this sense, that other forms of energy besides chemical energy undergo increase and decrease as the distribution of matter changes, and the heat evolved or absorbed by the reaction is a complex quantity, representing not merely the changes of chemical energy, but the resultant of the sum total of energy changes of all kinds. A *purely* chemical change is a fiction of much the same order as an *absolutely* rigid bar, or a *perfectly* frictionless constraint.

Further, it can be shown that in incomplete reactions (*i.e.*, in cases of equilibrium) the chemical energy of the system suffers no decrease during the course of the reaction; hence, if purely chemical energy were the only form of energy to be considered, it follows that all reactions leading to the establishment of an equilibrium ought to proceed without any heat absorption or evolution. But this is by no means the case, and so we are forced to the conclusion that in the so-called chemical changes other forms of energy besides chemical energy (*e.g.*, volume energy, heat energy, surface energy, &c.) must undergo transformations in order that the actually observed thermal disturbances may be accounted for.

But recognising the multiplicity of the specific energy changes which give rise to the thermal phenomena attendant on chemical change, the question may still be put, is there absolutely no connection under any conditions between the value and sign of the thermal changes and the necessity for the occurrence of chemical change? Cannot we predict under any circumstances which of two or more possible changes will occur?

It was a great advance that Horstmann made when he showed that the formulæ of thermodynamics can be applied to problems in the domain of chemistry. The leading features of the dissociation of such a substance as chalk, in accordance with the symbolism

$$CaCO_3 \leftrightharpoons CaO + CO_2,$$

are very similar to those of the evaporation of a homogeneous liquid such as water. But the application of the second law of thermodynamics to the process of evaporation leads to the establishment of very important relationships expressible in a simple formula, and Horstmann, on the strength of the strong analogy between the processes of evaporation and dissociation, applied this particular formula to certain cases of dissociation. The results to which the formula led were found to be in harmony with the facts; and it was not long before Horstmann showed that the principles and formulæ of thermodynamics generally are applicable to *all* cases of chemical equilibrium. In other words, the relations of thermodynamics and chemistry are not limited to the application of *one* special thermodynamical formula to a certain circumscribed class of equilibria distinguished as dissociations (see note 2, p. 199).

The verdict passed on Berthelot's " law " by thermodynamics, is that only at absolute zero would such a law obtain—that only at the unattainably low temperature of $-273°$ C. could prophecies founded on such a law be relied on. For, presupposing the possibility of chemical change at this temperature, it appears that at absolute zero all reactions would be complete; all reactions would take place with an evolution of heat, and of two or more possible reactions that one would occur which is attended with the greatest heat evolution. The higher the temperature rises above the absolute zero, however, the more does Berthelot's law " deviate from the truth inclining towards it." With the rise of temperature above zero enters, according to thermodynamic deduction, the possibility of incomplete reactions—of mobile equilibria, and, moreover, the higher the temperature the more does a given equilibrium shift in favour of the endothermic partial change.

Thus, the displacement of HNO_3 from $NaNO_3Aq$ by H_2SO_4Aq is endothermic; the displacement of H_2SO_4 from Na_2SO_4Aq by HNO_3Aq is exothermic. Hence it follows that a rise in temperature shifts the equilibrium of the system $NaNO_3Aq$, H_2SO_4Aq, HNO_3Aq, Na_2SO_4Aq in favour of the darker arrow as shown in the following symbolism.

$$Na_2N_2O_6Aq + H_2SO_4Aq \rightleftarrows Na_2SO_4Aq + H_2N_2O_6Aq.$$

Or, writing the reaction in equational form,

$$Na_2N_2O_6Aq + H_2SO_4Aq = x\,Na_2SO_4Aq + x\,H_2N_2O_6Aq +$$
$$(1-x)\,Na_2N_2O_6Aq + (1-x)\,H_2SO_4Aq,$$

we may state the same relation in another way by saying that x decreases in value as the temperature rises.

In cases where equilibrium is attained without thermal change, temperature is without influence thereon. Thus, nitric acid displaces hydrochloric acid from salt solution without evolution or absorption of heat; the same is true for the displacement of nitric acid from sodium nitrate solution by hydrochloric acid, hence in the equation

$$NaNO_3Aq + HClAq = x\,NaClAq + x\,HNO_3Aq +$$
$$(1-x)\,NaNO_3Aq + (1-x)\,HClAq$$

the value of x is independent of the temperature.

The equilibria resulting from the etherification of alcohols by acids (see p. 202) are also attained without thermal disturbance, and are accordingly found to be independent of temperature changes.

Having once and for all laid low the spectre of the "Law of maximum work," thermodynamics replaced it with a general law which, though of somewhat the same form as Berthelot's, differs from it in being universally true at all temperatures. This law, called the law of entropy, may be looked on as a particular form of statement of the Protean second law of thermodynamics. It states that a system is in stable equilibrium only when its entropy is a maximum, and therefore that any change which entails an increase in the entropy of the system is not only capable of spontaneous occurrence, but will in fact actually proceed until the entropy of the system reaches the maximum value attainable under the conditions.

Unfortunately this quantity or function entropy is a very difficult one to conceive, let alone measure, and the law concerning it is to chemists of more theoretical interest than practical use. We can, however, arrive at some kind of a conception of entropy by the following considerations.

The electrical energy of an isolated charged body is equal to the product $\frac{1}{2}QV$, where Q stands for the quantity of electricity with which the body is charged, and V is its potential. Suppose that for one body the electric energy is $\frac{1}{2}Q'V'$, and for another body $\frac{1}{2}Q''V''$. What are the conditions that electric energy shall pass from one body to the other? Simply electric connection and the inequality of V' and V''. If $V' > V''$, then electric energy will pass from the first body to the second until the potentials of both bodies is the same, and this will happen even if the electric energy of the first body is greater in amount than that of the second body. The passage of electric energy from one system to another is independent of the quantities of energy possessed by the two systems, and depends solely on their potentials. Electric energy always flows from places of high, to places of low potential.

On this account V is regarded as the "intensity factor" of electrical energy, and Q as the "capacity factor." Energy of other forms can similarly be resolved into capacity and intensity factors. For instance, the intensity factor of kinetic energy is velocity,[1] the capacity factor is mass, and so on.

What then are the factors of thermal energy? The intensity factor is very familiar to us and easily capable of evaluation—it is temperature. The capacity factor does not appeal directly to our senses, nor can we easily form a clear conception of it. Yet this factor, which we are content to define rather than conceive, has received the name entropy. Hence we can restate the law of entropy in the following way:— Any change which can increase the value of the capacity factor of the heat energy of a system takes place with readiness, and the system only then attains a position of stability (of disinclination to undergo any further change) when the value of this capacity factor is at a maximum.

In some forms of energy it is not possible to alter the capacity factor by adding energy to the system.

[1] Strictly speaking (velocity)^2, the kinetic energy of a mass m moving with velocity v being $\frac{1}{2} mv^2$.

The addition of kinetic energy to a moving bullet, for instance, cannot alter the mass of the bullet, which here represents the capacity factor. But in the case of heat energy it is otherwise; here it is possible, by adding thermal energy to a system, to alter thereby the thermal capacity factor of that system.

Suppose we have a mass of a solid substance S at the absolute temperature T. Let us impart to S an additional quantity of heat energy q under conditions such that the intensity factor of the heat energy of the system is not thereby altered, i.e., such that T remains constant. This may be simply realised by supposing T to denote the melting point of S, and q to be the latent heat of fusion of S. Let us call the first state of the system A and the final state B. To bring the system to the state A from absolute zero requires an amount of heat energy, say Q; to bring the system from zero to the state B requires $Q+q$ units of heat energy. Therefore in the passage from A to B the capacity of the system for heat energy has evidently increased, and the increase of heat capacity under the conditions named is called an increase of entropy, and is measured by the quotient $\frac{q}{T}$.

But thermodynamics has furnished us with tests of stability involving functions other than entropy; and some of these tests possess the advantage of being more easily applied to actual cases than the entropy test.[1]

---

[1] For details the reader is referred to Parker's *Elementary Thermodynamics*, also to Ostwald's *Lehrbuch der allgemeinen Chemie*, to which the author is indebted for the treatment of the subject of entropy given in the text.

A momentary glance at the expression defining the function known as the *free energy*, or *the thermodynamic potential at constant volume*, throws clear light on the question of Berthelot's "law." In the equation

$$F_A = U_A - T_A S_A$$

F_A denotes the free energy and U_A the total energy of a system in the state A. Its temperature in this state is T_A and its entropy S_A.

Let us consider another state, B, of this system, for which the values are F_B, U_B, T_A, and S_B, the volume and temperature of the systems remaining the same in both states.

From the two equations

$$F_A = U_A - T_A S_A$$

and

$$F_B = U_B - T_A S_B$$

we get by subtraction—

$$(F_A - F_B) = (U_A - U_B) - [T_A(S_A - S_B)].$$

Now, it results from thermodynamic reasoning that the change from the state A to the state B can only proceed of itself when the free energy of the system is decreased by the passage, *i.e.*, $(F_A - F_B)$ must be a positive quantity.

Rearranging the equation we obtain

$$(U_A - U_B) = \text{positive quantity} + T_A (S_A - S_B).$$

The value of S_B being greater than that of S_A, the whole term $T_A (S_A - S_B)$ must be negative.

Suppose the absolute magnitude of this term to be less than the "positive quantity" term, then $(U_A - U_B)$ must obviously have a positive value; that is to say, the change of the system from the state A to the state B has been attended by a loss of energy to the system —has been an exothermic change. But this is not the only possible case. If the absolute magnitude of the term $T_A(S_A - S_B)$ be greater than that of the positive quantity term, then $(U_A - U_B)$ must necessarily have a negative value; that is to say, the total energy of the system has been increased by its spontaneous passage from the state A to the state B—the change has been an endothermic one.

It is clear that the importance of the term $T_A(S_A - S_B)$ rises *pari passu* with the temperature at which the change proceeds. In other words, the higher the temperature the more likely are endothermic reactions to occur.

THE END.

PRINTED BY BALLANTYNE, HANSON AND CO.
EDINBURGH AND LONDON.

www.ingramcontent.com/pod-product-compliance
Lightning Source LLC
Chambersburg PA
CBHW030111030726
47498CB00007B/2341